Automated Vehicles Symposium 2014

Drivers. Vehicles. Infrastructure.

San Francisco, California, USA
15 – 17 July 2014

ISBN: 978-1-5108-5181-8

Printed from e-media with permission by:

Curran Associates, Inc.
57 Morehouse Lane
Red Hook, NY 12571

Some format issues inherent in the e-media version may also appear in this print version.

Printed by Curran Associates, Inc. (2018)

Association for Unmanned Vehicle Systems International
2700 South Quincy Street
Suite 400
Arlington, VA 22206
USA

Phone: (703) 845-9671
Fax: (703) 845-9679

www.auvsi.org

Additional copies of this publication are available from:

Curran Associates, Inc.
57 Morehouse Lane
Red Hook, NY 12571 USA
Phone: 845-758-0400
Fax: 845-758-2633
Email: curran@proceedings.com
Web: www.proceedings.com

Acknowledgments

The 2014 Automated Vehicles Symposium (AVS 2014) was produced through a partnership between the Association of Unmanned Vehicle Systems International (AUVSI) and Transportation Research Board (TRB), and is indebted to its many volunteer contributors and speakers who developed the symposium content, produced sessions, and contributed to the development of these proceedings. AVS 2014 also benefitted from the direct financial contributions of its commercial benefactors, and the institutional support of the many companies and agencies that supported the time, travel, and participation for their staff. The organizing committee gratefully acknowledges the United States Department of Transportation (U.S. DOT) Intelligent Transportation Systems Joint Program Office (ITS JPO) for the level of its support to AVS 2014 and for these proceedings.

Symposium Executive Committee

David Agnew, Continental Automotive| Member, AUVSI Board of Directors

Richard Bishop | AUVSI subject matter expert on automation

Richard Cunard | Senior Program Officer, Traffic and Operations Engineer, TRB

Bob Denaro, Independent Consultant | Chair, TRB Joint Subcommittee on the Challenges and Opportunities for Road Vehicle Automation

Jane Lappin, Volpe National Transportation Systems Center | Chair, TRB Intelligent Transportation Systems Committee (AHB15)

James Misener, Qualcomm | Symposium Demonstrations Coordinator

Steven Shladover, University of California PATH Program | Chair, TRB Vehicle-Highway Automation Committee (AHB30)

Michael Toscano | President and CEO, AUVSI

Lindsay Voss | Senior Program Development Manager, AUVSI

Benefactors

AutonomouStuff — Today's Products for Tomorrow's Solutions

BOSCH — Invented for life

here — Maps for Life

Continental

KVH

DENSO

MAGNA — MAGNA ELECTRONICS

TRAVELERS

Velodyne LiDAR

Breakout Session Organizers and Symposium Volunteers

Osman Altan, Federal Highway Administration
Carl Andersen, Federal Highway Administration
James Anderson, RAND
Bart van Arem, Delft University of Technology
Chris Armstrong, Leidos
Thomas Bamonte, North Texas Toll Authority
Robert Bertini, Portland State University
Chandra R. Bhat, University of Texas at Austin
Paul Carlson, Texas A&M University
Ching-Yao Chan, University of California Berkeley PATH
Nick Cohn, TomTom
John Collins, Transportation Business Law and Strategy
Janet Creaser, University of Minnesota
Tinu Diver, U.S. DOT/Volpe Center
Kevin Dopart, U.S. DOT/ITS JPO
Frank Douma, University of Minnesota
John Estrada, Dering & Estrada
Dan Fagnant, University of Utah
Robert Ferlis, Federal Highway Administration
Greg Fitch, Virginia Tech Transportation Institute
Maxime Flament, ERTICO
Erin Flanigan, Cambridge Systematics
Edward Fok, Federal Highway Administration
Deborah Freund, Federal Motor Carrier Safety Administration
Kevin Gay, U.S. DOT/Volpe Center
Ryan Gerdes, Utah State University
Anna Giragosian, Leidos
Ginger Goodin, Texas A&M Transportation Institute
Tom Healy, National Highway Traffic Safety Administration
Kevin Heaslip, Utah State University
Charlie Howard, Puget Sound Regional Council
Reuben Juster, University of Maryland
Keith Kahl, Oak Ridge National Laboratory
Ram Kandarpa, Booz Allen Hamilton
Valarie Kniss, U.S. DOT/Volpe Center
Walt Kulyk, Federal Transit Administration
Alain Kornhauser, Princeton University
Scott Le Vine, Imperial College London
Rachel Liu, New Jersey Institute of Technology
Elizabeth Machek, U.S. DOT/Volpe Center
Michael Manser, Texas A&M Transportation Institute
Yeganeh Mashayekh, University of Pennsylvania
George Munteau, Pacific Northwest National Laboratory
Suzanne Murtha, Atkins
Siva Narla, Institute of Transportation Engineers
Markos Papageorgiou, Technical University of Crete
Ellen Partridge, U.S. DOT
Scott Paxton, Bowman and Brooke
Ram Pendalya, Arizona State University
Joseph Peters, Federal Highway Administration
Mohammad Poorsartep, Texas A&M Transportation Institute
Lydia Rainville U.S. DOT/Volpe Center
Brian Routhier, Federal Motor Carrier Safety Administration
Shelley Row, Shelley Row Associates, LLC
Shannon Sanders McDonald, Southern Illinois University Carbondale
Chris Schwarz, University of Iowa
Robert Seidl, Motus Ventures
Nazy Sobhi, U.S. DOT/Volpe Center
Karlyn Stanley, RAND
Francine Steelman, Miami-Dade Expressway Authority
Thomas Stephens, Argonne National Laboratory
John Suh, Hyundai Ventures
Sudharson Sundararajan, Booz Allen Hamilton
Joan Walker, University of California Berkeley
Eric Williams, Insurance Institute for Highway Safety
Steve Wood, National Highway Traffic Safety Administration
Stanley E. Young, University of Maryland
Wei-Bin Zhang, University of California Berkeley PATH
Ismail Zohdy, Booz Allen Hamilton

Contents

1 Overview of Symposium

The 2014 Automated Vehicles Symposium (AVS 2014) was organized and produced through a partnership between the National Academies of Science and Engineering Transportation Research Board (TRB) and the Association for Unmanned Vehicle Systems International (AUVSI) to serve their shared constituencies' interests in understanding the impact, benefits, challenges, and risks of increasingly automated road vehicles and the environments in which they operate. AVS 2014 built on the 2013 TRB Workshop on Vehicle Automation and the AUVSI 2013 Driverless Car Summit to bring together key government, industry, and academic experts from around the world with the goal of identifying opportunities and challenges and advancing automated vehicle (AV) research in a range of disciplines.

The symposium took place over five days, 14-18 July, and included three days of core activities and ancillary sessions on the first and last days. Section 1.2 presents an overview of the program. The morning plenary sessions included presentations from the public sector, automakers and suppliers, and academic research institutes. The afternoon sessions consisted of 10 breakout sessions on the following topics:

1. Evolutionary and Revolutionary Pathways to Automated Transit and Shared Mobility
2. Regional Planning and Modeling Implications of Driverless Cars
3. Roadway Management and Operations with Automated Vehicles
4. Truck Automation Opportunities
5. Legal Accelerators and Brakes
6. The State and Future Direction of Automated-Vehicle Human Factors
7. Near-Term Connected/Automated Technology Deployment Opportunities
8. Personal Vehicle Automation Commercialization
9. Technology Roadmap, Maturity and Performance: Operational Requirements for Vehicle-Road Automation Systems and Components
10. Road Infrastructure Needs of Connected-Automated Vehicles

The workshop also included three ancillary sessions:

- Society of Automotive Engineers (SAE) International On-Road Automated Vehicle Standards (ORAVS) Committee
- U.S. DOT Listening Session
- Friday Ancillary Workshop: Envisioning Automated Vehicles within the Built Environment: 2020, 2035, 2050

This synopsis provides a brief summary of highlights from all symposium activities, including plenary session presentations, ancillary sessions, and demonstrations. Where available, links to symposium materials and presentations are embedded in this synopsis.[1] The body of this document includes a brief summary of each breakout session's focus and goals, agenda, key issues and results, and next steps. Complete breakout session proceedings for each group are provided by courtesy of the breakout session organizers as appendices.

[1] All links are current as of September 15, 2014

1.1 Symposium Attendees

Over 570 attendees participated in the symposium over the course of the week. Attendees represented a wide range of organizations from government and industry to the academic-, public-, and private-sector research communities (summarized in Figure 1 below). These participants represented disciplines ranging from engineering to psychology to law. International participants attended from 15 countries.

Figure 1: (Left) Attendees gather at the 2014 Automated Vehicles Symposium (Image Source: AUVSI); (Right) Summary of attendee affiliations and home state and country (Source: David Agnew, Continental Automotive, AUVSI Board of Directors).

1.2 Workshop Agenda

	Morning	Afternoon	Evening
Monday 7/14		Automated Vehicle Standards & Best Practices: SAE On Road Automated Vehicle Standards Committee	
Tuesday 7/15	Plenary Sessions Opening Keynote Address Vehicle Manufacturers and Suppliers Briefings Panel Session: Digital Infrastructure, Technology Challenges	Breakout Sessions	Networking Reception and Poster Presentations
Wednesday 7/16	Plenary Sessions Vehicle Manufacturers and Suppliers Briefings Automation Project Updates from Europe Panel Session: Societal Issues and Non-technical Challenges	Breakout Sessions	Networking Reception and Poster Presentations
Thursday 7/17	Plenary Sessions Public Sector Addresses Breakout Session Reports	U.S. DOT Automation Research Presentations and Listening Session	
Friday 7/18	Ancillary Workshop Envisioning Automated Vehicles within the Built Environment: 2020, 2035, 2050		

1.3 Demonstrations

Another component of the workshop was a combination of on-road and controlled area demonstrations. James Misener (Qualcomm) was lead organizer for the demonstrations with support from other organizers. The location of the symposium was chosen for its proximity to the highway and its parking lot to enable both highway and static demonstrations. A description of each of the demonstrations is below.

Velodyne's LiDAR division demonstrated what it's like to look through the "eyes" of a self-driving car, with its HDL-64E 3D, real-time LiDAR sensor.

Peloton displayed two Class 8 tractor trailers equipped for Level 1+ automation (left) while Cruise Automation displayed its RP-1 highway autopilot prototype (right)

AutonomouStuff provided an interactive vehicle demonstration presenting PolySync™, The Autonomy Operating System.

Quanergy demonstrated a 3D sensing system based on its compact low-cost high-performance high-reliability Mark VIII LiDAR. The demo featured a live display showing the surroundings with objects that are detected, tracked, and classified in real time.

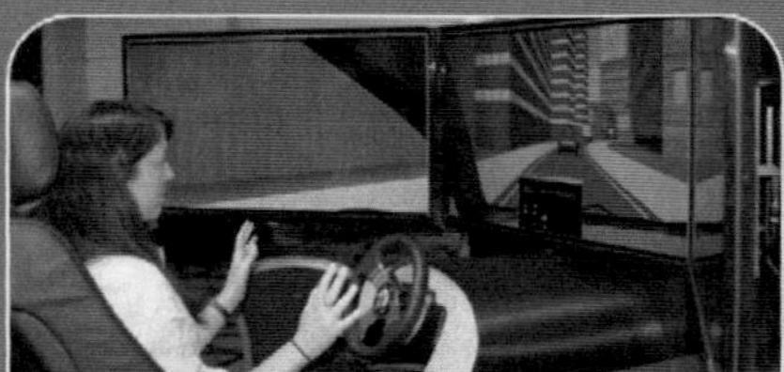

Magna demonstrated a monocular camera-based Traffic Jam Assistance system, capable of automatically operating in stop and go traffic at speeds up to 44 mpg (bottom). Magna also demonstrated this same system with the use of a simulator inside the symposium (top).

Demonstration Image Source: Volpe

2 Overview of Plenary Sessions: Day 1

This section of the synopsis contains a brief overview of plenary sessions from the first day of the symposium, including the speaker, selected highlights from each presentation and panel discussion, with a link to the presentation slides if available. These summaries were drawn from presentation materials and symposium notes.

2.1 Opening Keynote

PRESENTATION	KEY HIGHLIGHTS
Opening Keynote Address: The Promises and Pitfalls of Vehicle Automation Ralf Herrtwich, Director of Driver Assistance and Chassis Systems, Research and Advanced Engineering Group, Daimler Automotive Group	• In the near-term, the focus of early automation is to enhance existing products; longer-term automation may ultimately reshape and transform mobility. • Level 2 is arguably the safest form of automation if drivers only paid attention. The driver is in the loop and can supervise the system. The problem is that drivers are increasingly distracted. • Near-term approach: limiting exposure to automation in certain operating conditions (e.g., roads, speeds, weather). • Studies show that in Germany the mean distance between accidents is 7.5M kilometers. While automated vehicles may be able to eliminate many of those accidents, how do we ensure that they can match the impressive performance of humans for the intervening 7.5M kilometers? • If a human driver injures a pedestrian while maneuvering in a complex environment, it is an accident, while if an automated vehicle injures a person it was programmed to take that action as a response to the complex environment. What are the legal implications of that? • Safe driving is not just about crash avoidance, it is about obeying traffic rules and understanding when a red light is coming up ahead. • There is a need for a validation debate – how do we ensure a vehicle is safe in arbitrary traffic? Practically is it impossible to test all use cases for an AV. • Level 4 autonomy features such as automatic valet would change the economics of car sharing services.

2.2 Briefing: Vehicle Manufacturers and Suppliers

PRESENTATION	KEY HIGHLIGHTS
Jan Becker, Director, Automated Driving Engineering, Robert Bosch, LLC. Slides	• The future mobility context is automated, connected, and electric. • Automated driving starts with highway driving and parking functions. A step-by-step approach is beneficial for technological and psychological reasons. • Prerequisites for automated driving are: surround sensing, safety and security, global standards, clear liability, dynamic map data, and system architecture. • Developing automated driving functions calls for profound knowledge and changes for all vehicle systems (e. g. sensors, actuation, electrical/electronic architecture, semiconductors, driver monitoring, and automotive cloud). • Bosch believes that LiDAR is required as a third sensor to mitigate the environmental shortcomings of currently used radar and video sensors. • Highly automated driving requires the latest high-precision map data with aggregated information processing and delivery via the cloud which is not likely to be achieved by a single company. • Expenditure for validation could increase by a factor of 10^6 to 10^7. Highly automated systems require completely new release strategies. • Bosch has all the necessary key technologies available and is getting them ready for market entry.
Cris Pavloff, Advanced Technology Engineer, Technology Office, BMW Group Slides	• Highly automated driving will increase safety, comfort, and efficiency for the driver and the traffic system. • Cooperative Intelligent Transportation Systems (ITS) is an essential part of our architecture for highly automated driving, providing information beyond the 200 meter sensor range; this places demands placed on its accuracy, availability and penetration. • Physical and digital infrastructure such as dynamic traffic management, real-time traffic information, maps and landmarks, lane markings, and signage are all important to highly automated driving. • The success of highly automated driving requires agreement on legal and technical standards, regulatory framework, and defining insurance models. • Currently working with Continental to create and test an electronic copilot system; expanded testing is expected in 2018 with potential global rollout in 2020.

PRESENTATION	KEY HIGHLIGHTS
Maarten Sierhuis, Director, Nissan Research Center Silicon Valley	• The Nissan Q50 has the first production drive-by-wire system—a first and necessary step towards automation. • Current testing includes a 360-degree simulator of roads and challenging intersections around their research facility. • The challenges in human machine interface and interaction (HMII) extend to the vehicle architecture. • Human machine interaction is a particularly difficult challenge to solve. The research falls into three categories: transition to and from autonomous mode, interaction between human and vehicle, and communication with people outside the vehicle. • Humans create mental models, but the system does not have a mental model of the driver. Should it? How do we avoid a fight? • It is important to understand the cultural aspects of vehicle use – ethnography of driving in daily life.
Manufacturers and Suppliers Briefings Panel Question & Answer Moderator—Bob Denaro	• Lost connectivity: o Connectivity would enhance other sensing and awareness. In its absence the vehicle would have to fall back on on-board sensors. o Industry needs to solve the problem without connectivity first. • Base map standards and requirements: o This issue is still being worked out. A challenge is that the more information in the map, the less reliance on sensors is required; however, real-time accuracy requires continual and dynamic updates. • Automation in severe weather: o Researchers are currently working on improving sensors to work at night, in rain, and in snow, as well as self-cleaning glass and paint. o Materials development is another aspect of this research. • We may need a shift in expectations to automation as a service that may not be available all of the time, rather than a piece of hardware.

2.3 Panel Session: Digital Infrastructure

PRESENTATION	KEY HIGHLIGHTS
Ogi Redzic, Vice President, Connected Driving, HERE/Nokia Slides	• Through High Definition (HD) Maps, Live Roads, and Humanized Driving, the HERE Automated Driving Framework effort underway addresses three problems: Where exactly am I? What lies ahead? How can I get there comfortably? • HD Maps will help vehicles properly locate themselves by providing a 3D supply model and surface, with information on road surfaces, lanes, lane boundaries, and proper maneuvers. • The critical piece for HD maps is a continuous, dynamic update loop. • HD Live Roads would use connectivity and layer dynamic information on top of the HD map to provide drivers and vehicles with the ability to plan beyond sensor visibility. • Conducted research using probe data to study exit ramp speed profiles. This type of study informs Humanized Driving, which will make automation more appealing during adoption.
Andrew Chatham, Principal Engineer, Self-Driving Cars, Google[X]	• Defined a map broadly: geographic information provided to the vehicle in advance. This information may be a physical or invisible, such as an intersection, implied crosswalk, or speed limit. • Noted the importance is on consistency, rather than precision. • The Google[X] vehicles use some physical infrastructure as input to automated driving. • Efforts to date have focused on freeway driving, but Google[X] is now refocusing on street driving. • Through a semi-automated process, the operations team is adding a logical layer to a graphical map (e.g., driveways, which lanes are controlled by which traffic signals). • Vehicles should be robust to changes in the world; maps make driving easier, but they can't be relied on 100% of the time. • Agencies currently use their own unique formats in construction databases. Some kind of standardization will be needed.

PRESENTATION	KEY HIGHLIGHTS

Digital Infrastructure Panel Question & Answer
 	
Moderator—Bob Denaro
	• Accuracy of sensors for developing HD maps:
	  o Current in-vehicle sensors are not precise enough, but improvements are ongoing.
	  o Aerial LiDAR may help for mapping furniture, but for the road there is no substitute for driving and collecting the data.
	• Lane center lines and markings:
	  o Both HERE and Google are moving to recording where the lane markings are and other features rather than depicting a road as an assumed centerline in the past. Google now maps lanes in a way that defines lane parameters rather than lane centers, as with road edges and lane separation markings.
	• Variability in lane marking exists within and across countries. There has to be some level of ability in the vehicle to tolerate this.

2.4 Panel Session: Technology Challenges

PRESENTATION	KEY HIGHLIGHTS
Alberto Broggi, Vislab, University of Parma Slides	• Focused on challenges of physical perception including estimating and measuring distances and speeds and recognizing road players. • No single technology can cover all of the challenges. It is important to try to evaluate uncovered scenarios. An open question is how many different and overlapping technologies are needed. • Continued research and development to extend each sensor's capabilities and try to understand how to improve them is essential. • Every sensor should be stressed to the limit so there are realistic expectations about applicability. • Research to create a fusion system to get one single set of sensors that can deliver necessary information is essential.

PRESENTATION	KEY HIGHLIGHTS
Markos Papageorgiou, Director, Dynamic Systems & Simulation Laboratory, Technical University of Crete Slides	• Intelligent vehicles may lead to "dumb" traffic flow if not managed appropriately. • The effects of high penetration levels of automated driving on traffic management have not yet been considered. • If starting tomorrow every vehicle were automated, the result would be AV gridlock. For example, AVs may stop under certain circumstances, and adaptive cruise control (ACC) has negative impacts on congestion, plus sluggish acceleration. • Cooperative systems are key, but they may introduce other challenges. For example, cooperative adaptive cruise control (CACC) platoons could generate challenges for lane changing, merging, and exiting. • We will need to implement cooperative traffic management, with consideration for objective, complexity, architecture, actuators, compatibility, and transition period.
John Leonard, Professor of Mechanical and Ocean Engineering, Massachusetts Institute of Technology Slides	• "It's hard to convey to the public how hard this is." Major unsolved issues exist; caution must be exercised in not overhyping how well automated driving works. • Automated driving is a transformative technology that can/will change the world, but there are many open questions. • Interpreting human gestures and social cues will be some of the most challenging communications problems. • Other open technical challenges include: o Left-turn across high-speed traffic onto busy roads; o Effect of changes in road surface appearance on map-based localization; o Capability to predict what will happen next in demanding conditions; o Operations in adverse weather.

PRESENTATION	KEY HIGHLIGHTS
Michael Wagner, Carnegie Mellon University	• Automation brings robots into people's daily lives. We need to be able to answer: Do we trust these systems? Are we confident that they have been developed carefully? • Serious risks can be posed by software, including code defects or even more perniciously, weaknesses of architectures and design. • The software that controls automated vehicles suffers from every kind of complexity: more code, more conditional branches, high-dimensional interfaces, and complex (often novel) algorithms. • We must consider a wide array of traditional and nontraditional security concerns. Because these systems are physical the nature of what makes a security exploit can be different than the kinds traditionally faced by information technology systems. • Traditional testing and standards are insufficient for justifiably instilling trust in the complex software that drives driverless cars. • Subjecting software to unexpected input combinations is an efficient way to find weaknesses in robustness and security. • Runtime verification, an approach that applies formal logics, has a role to play in many applications.
Technology Challenges Panel Question & Answer Moderator—Steven Shladover	• Legality of the Vislab on-road driverless experiment in Parma, Italy: o Vislab worked with local transportation and law enforcement stakeholders for authorization for that specific test. • Key breakthrough technology challenges for automation: o Dynamic environment mapping o Based on testing in the Southwest United States, one manufacturer described a sandstorm use case as an edge case for hardware sensors o Sensors that measure variable exactly and provide Immediate feedback of failure • Scalable traffic management methodology and architecture.

2.5 Automated Vehicles Symposium Attendee Survey Results

PRESENTATION	KEY HIGHLIGHTS
Steve Underwood, University of Michigan, Dearborn	• The survey objective was to explore the range of opinions among AVS 2014 attendees on the future of AV functions and availability. • Top 3 barriers cited were legal, regulatory, and cost. • Equal number of respondents rated technology highest and lowest. • For level of safety compared to today, 56% said as-safe to 2x as-safe; 36% said 10x to perfectly safe • 73% said society will accept some automation-caused accidents. • 46%/54% said practical/not practical to expect driver to respond when needed • 67% said that vehicle-to-vehicle (V2V) is essential for full automation • While the mean date predicted for full automation of a taxi in all conditions was 2030, when separately asked when they would let their children or grandchildren ride in a fully automated taxi alone to school, approximately 25% of respondents cited 2040 or later a nearly 8% said never.

Dr. Steve Underwood from the University of Michigan-Dearborn administered an online survey to registrants in advance of the symposium regarding their opinions on the likely deployment forecasts for automated vehicles. The charts below summarize several key results based on 220 responses.

The timeline at right represents survey respondents' opinions of likely deployment years for vehicle automation systems falling within SAE levels 3, 4, and 5.

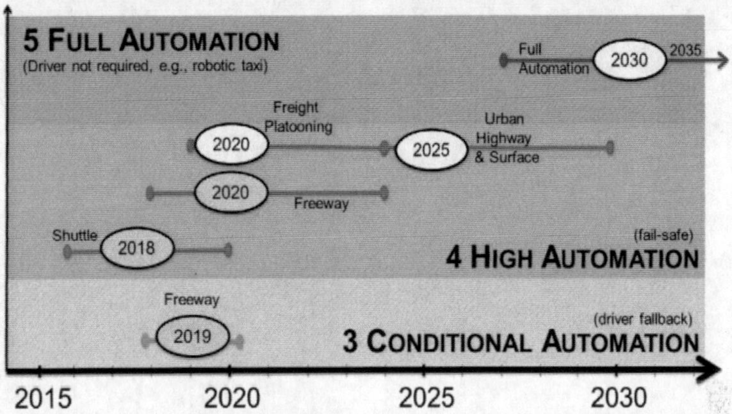

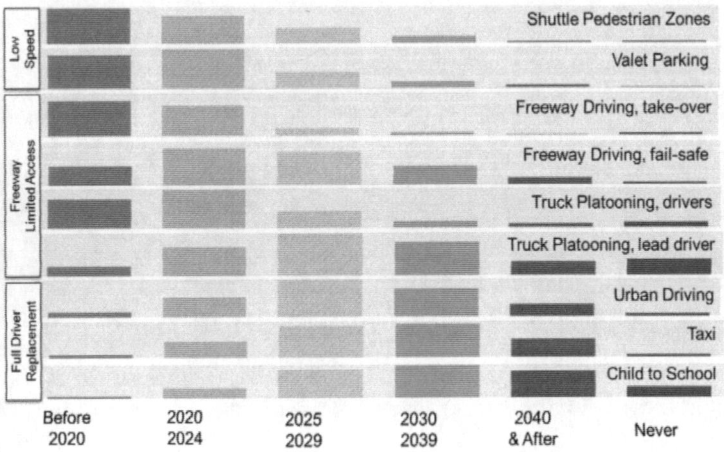

Distributions of responses among AVS 2014 attendees regarding the likely deployment years for vehicle automation systems falling within SAE levels 3, 4, and 5.

Figure 2: Results from a survey administered by the University of Michigan's Dr. Steve Underwood among AVS 2014 attendees focused on likely deployment timelines for automated vehicle systems (Image Source: Steve Underwood).

Figure 3: Dr. Ralf Herrtwich, from Daimler AG, delivers the opening keynote address at the 2014 Automated Vehicles Symposium (Image Source: AUVSI).

Figure 4: Michael Toscano, President and CEO of AUVSI, and Jane Lappin, U.S. DOT Volpe Center, Chair of the TRB Intelligent Transportation Systems Committee, welcome attendees to the 2014 Automated Vehicles Symposium in San Francisco, CA (Image Source: Volpe Center).

Figure 5: Andrew Chatham, Principal Engineer on the Google[X] Self-Driving Car Project, discusses the digital mapping systems that enable the self-driving capabilities of the company's prototype vehicles (Image Source: Volpe Center).

Figure 6: (Left to Right) Michael Wagner, from Carnegie Mellon University; Dr. John Leonard, Professor of Mechanical and Ocean Engineering at the Massachusetts Institute of Technology; Dr. Markos Papageorgiou, Director of the Dynamic Systems & Simulation Laboratory at the Technical University of Crete; and Dr. Alberto Broggi, from the Vislab and University of Parma participate in a panel focused on automated vehicle technology challenges at the 2014 Automated Vehicles Symposium (Image Source: Volpe Center).

3 Overview of Plenary Sessions: Day 2

This section of the synopsis contains a brief summary of plenary sessions from the second day of the symposium, including the speaker, selected highlights from each presentation and panel discussion, with a link to the presentation slides if available. These summaries were drawn from presentation materials and symposium notes.

3.1 Keynote Address

PRESENTATION	KEY HIGHLIGHTS
Clifford Nass Memorial Lecture – The Human Side of Automation Don Norman, Director, Design Lab, University of California (UC) San Diego and Author of "The Design of Future Things"	• Incorporate human factors considerations from day one. • We want collaboration between the person driving and the technology driving. • People are creative, flexible, imaginative, and handle unexpected events. Machines like control, accuracy, and consistency. • If 95% of accidents are attributed to human error, it means that humans are being asked to do things they are not good at. • Noted that aviation is still not fully automated. • A need exists to develop more natural ways of communicating, such as haptic warnings for lane departure. • "Technology is designed for people. It isn't enough to be a technologist, you have to understand people." • I am in favor of full automation; the problem is when we have partial automation.

3.2 Briefing: Vehicle Manufacturers and Suppliers

PRESENTATION	KEY HIGHLIGHTS
John Capp, Director, Electrical & Controls Systems Research & Active Safety Technology Strategy, General Motors (GM) Slides	• Demonstration vehicles have created a perception that we are on the verge of transformation of how cars are built; the reality is more gradual. • The Defense Advanced Research Projects Agency (DARPA) Urban Challenge was a catalyzing event for automation. • For automation to succeed, it is essential that the product must be something that customers want to buy. • An integrated, advanced driver assist package is already available for Cadillac. Supercruise will be the next step in the automation path for GM. • Human factors studies are important for revealing unintended consequences. The complexity of these studies will increase as the technology advances. • Technology enablers and advancements are needed for higher levels of automation: perception and algorithms, advancements in 360-degree sensing, sensor fusion, maps, global positioning system (GPS), and V2V/vehicle-to-infrastructure (V2I) integration.

PRESENTATION	KEY HIGHLIGHTS
Steffen Linkenbach, Director, Systems & Technology, North American Region, Continental Slides	• The absolute basis for the successful introduction of automated driving is driving safety. Dynamics, efficiency, and comfort are the benefits that will result from a solid foundation that is built on safety. • Continental conducted an extensive study to gauge the attitudes, knowledge and level of acceptance for automated driving. • To ensure driving safety and customer acceptance, a phased-in, step-by-step approach is essential to the success of automated driving. • The step-by-step introduction of technology should help ensure that the technology costs also are reduced along the way. • On average, the Americans surveyed would pay only four to five percent of the average price of a car. Right now, that is not a realistic expectation based on the costs of technology.
Patrice Reilhac, Innovation & Collaborative Research Director, Comfort & driving Assistance Business Group, Valeo Slides	• Conducted a global intuitive driving experience study and found: 1) Automated driving must be intuitive (automated + connected + extended human machine interface (HMI) 2) Support for automated driving starts from low speed onwards. • The study found automated parking, emergency braking, traffic jam, and highway automation to be attractive to customers. • The extended HMI linking the automated and connected car was the key to success of the park4u demonstration at the Frankfurt auto show. • Connectivity is a broad concept encompassing a number of technologies and networks ranging from the car network to short-, medium-, and long-range wireless.
Realizing Self-Driving Vehicles Chris Urmson, Director, Self-Driving Cars, Google[X]	• The Google[X] focus is shifting from highway driving towards neighborhood driving, consistent with their mission of freedom and mobility. • The value proposition is more than just safety, it is about better well-being. • Current efforts include translating the geometry of the world to a semantic meaning (e.g., a school bus is different than a big car and requires different behavior). • 140 Google employees currently use automated vehicles for their commute, making theirs the largest automated vehicle fleet in use. • Showed a series of videos of the Google automated vehicles in neighborhood conditions (e.g., interacting with cyclists, in a construction zone).

PRESENTATION	KEY HIGHLIGHTS

Vehicle Manufacturers and Suppliers Briefings Panel Question & Answer
 	
Moderator—Richard Bishop
	• Availability of automated parking:
	  o Parking is a variable function with different levels from assistance to full automation.
	  o Increased availability and functionality in the next months and years.
	  o There are already 3.5 million cars with driver-monitored automated parking (both parallel and home garage).
	• Shared mobility:
	  o Personal vehicles are a type of freedom; it is difficult to encourage sharing.
	• Role of crash avoidance when automation is off:
	  o Operating in the background

3.3 Updates on Automation Projects in Europe

PRESENTATION	KEY HIGHLIGHTS
CityMobil2: Automated Road Transport Systems in Urban Environments Adriano Alessandrini, University of Rome La Sapienza Slides	• Showed a live feed from Sardinia, Italy, the location of the first CityMobil2 demonstration. • The system is the first of seven large-scale demonstrations. • Pedestrians and cyclists share the corridor, which is separated from motorists. • The proposed certification procedure based on EN50126 takes advantage of "type approval" on motor vehicles directives. • A novel feature was analysis of locations where sensor line-of-sight is blocked by a wayside sign or object, and adjusting the infrastructure by erecting barriers to slow pedestrians or bicyclists to allow adequate lead time in detection by vehicle sensors. • Two contiguous but independent infrastructures likely to evolve: 1) automated road transport systems (ARTS) have dedicated or segregated lanes, with manually driven vehicles having access if rules are obeyed, and 2) conventional manually driven vehicles lanes which are not accessible to ARTS vehicles.

PRESENTATION	KEY HIGHLIGHTS
The AdaptIVe Project: Working on Research, Legal, and Deployment Issues in Europe for Automated Vehicles Angelos Amditis, Research Director, I-Sense Group, Institute of Communication and Computer Systems Slides	• A successor project to other European Commission projects such as HAVEit, interactIVe, and SARTRE that addressed advanced driver assistance systems (ADAS), cooperative safety systems, platooning, and automation. • The project includes several subprojects name that address legal issues, human-vehicle integration, automation in close-distance scenarios, automation in urban and highway scenarios, and evaluation. • The scope of application domain addressed in the program includes parking, urban, and highway assistants (SAE levels 1-4), is suited for mixed traffic, and provides adaptive support based on the driving task demand. • Some of the functions are: lane following, stop and go driving, cooperative merging, danger spot intervention, predictive automated driving, park assistant, city cruise, and city chauffeur. • Classification by level of automation and speed is not sufficient for further work. Response 4, one of the projects within AdaptIVe will develop and verify the selection of additional classification parameters from legal, functional safety, and human factors perspectives. • Legal aspects of Response 4 address international, national, liability, and data privacy and data security.
The Drive Me Project: Autonomous Driving by Volvo Anders Tylman-Mikiewicz, General Manager, Volvo Monitoring & Concept Center, Volvo Car Corporation	• Safety is the guiding principle for Volvo. • Automation would also reduce the complication in their customers' lives and allow them to maximize their time. • The $75 million project is the world's first pilot of 100 automated vehicles in real traffic with real customers. Planning is underway for a 2017 demonstration. • The project is a cooperative effort between Volvo Car Group, Lindholmen Science Park, the Swedish Transport Administration, the Swedish Transport Agency, and City of Gothenburg. • During the demonstration, the highway loop being used will be monitored and "certified" for automated operation. • The technology will include camera, radar, LiDAR, map data, and cloud-based communication. • Showed a DriveMe promotional video (video link)

3.4 Panel Session: Societal Issues and Non-Technical Challenges

PRESENTATION	KEY HIGHLIGHTS
Ginger Goodin, Texas A&M Transportation Institute Slides	• Forces unrelated to vehicle technology will dominate the state and local transportation policy focus in the near term. • State and local agencies are the ultimate implementers of policy changes and serve as policy innovation labs, so understanding their perspectives and motivations is key. • Uncertainty in AV capabilities, benefits and development path will contribute to the lack of clarity needed for policy development. • Unresolved public policy and institutional issues will play a role in slowing deployment. • The public sector case for automation will need a financial component. • Actions supporting state and local policy development include: Industry and government collaboration and communication, field demonstrations and pilot projects, and plausible deployment scenarios.
Michael Gucwa, Management Science and engineering Department, Stanford University Slides	• Presented findings from Ph.D. research seeking to answer the question: How will automation change the daily travel decisions of individuals and alter overall vehicle miles traveled (VMT) and energy use? • The analysis used Metropolitan Transportation Commission Travel Model One. • The scope and assumptions for the analysis were advanced level 3 automation, urban travel, and status quo for vehicle ownership and form. • Results showed a "most plausible" outcome of an expected short-run increase of 4-8% in daily vehicle miles travelled with automation. In the long run people could change other larger decisions. • Open questions include long-term land use adjustments, welfare and equity, the role of policy, level 4 automation, and shared ownership.
Ken Laberteaux, Toyota Research Institute North America Slides	• For over 100 years, each new commuting mode offering higher speed has increased commute distances in the United States. • Without policy changes, in the United States, levels 2 and 3 automated driving may likely increase highway speeds, VMT, commute distances, and accelerate suburbanization trends. • Increased speed offers home buyers a larger area in which to trade-off price vs. location amenities (e.g. public school quality).

PRESENTATION	KEY HIGHLIGHTS
Mike Van Nieuwkuyk, Executive Director of Global Automotive, J.D. Power and Associates Slides	• Presented results from the J.D. Power 2014 United States Automotive Emerging Technologies Study. • Interest in fully autonomous driving continues to rise slightly year-over year, as consumers are becoming more aware of or experiencing semi-autonomous features. • Because of a misalignment between consumer expectations and near-term technology capabilities, it will be important to set realistic expectations. • Interest in automated driving varies by gender, age, and other demographic factors.

Figure 2: (Left to Right) Patrice Reilhac from Valeo; Steffen Linkenbach from Continental; Chris Urmson, from Google [X], and John Capp from General Motors participate in a vehicle manufacturer and supplier briefing at the 2014 Automated Vehicles Symposium (Image Source: Volpe Center).

Figure 8: Adriano Alessandrini, from the University of Rome La Sapienza and Project Coordinator for the CityMobil2 project, discusses the CityMobil2 demonstration in Sardinia while his colleague demonstrates the system's capabilities via live video feed (Image Source: Volpe Center).

Figure 9: Don Norman, Director of the Design Lab at the University of California San Diego, delivers the Clifford Nass Memorial Lecture at the 2014 Automated Vehicles Symposium (Image Source: AUVSI).

4 Overview of Plenary Sessions: Day 3

This section of the synopsis contains a brief summary of plenary sessions from the third day of the symposium, including the speaker, selected highlights from each presentation and panel discussion, with a link to the presentation slides if available. These summaries were drawn from presentation materials and symposium notes.

PRESENTATION	KEY HIGHLIGHTS
U.S. Department of Transportation Address Speaker: Kevin Dopart, Program Manager for Connected Vehicle Safety and Automation, Intelligent Transportation Systems Joint Program Office, U.S. DOT (ITS JPO) Slides	• Connected Vehicles (CVs) are still a major focus of the U.S. DOT ITS research program and investment; however, automated vehicle research across all levels is ramping up. • The final Multimodal Program Plan for Vehicle Automation will depend on the ITS Strategic Plan and budget. • The U.S. DOT is collaborating domestically with other Federal agencies—the U.S. Department of Defense (DoD), Department of Energy (U.S. DOE), and the National Aeronautics and Space Administration (NASA)—as well as internationally through the Tri-Lateral Working Group on Automation in Road Transport. • The U.S. DOT roles in vehicle automation include: o Identifying benefit opportunities in automated vehicle technology; o Facilitating development and deployment of automated transportation systems that enhance safety, mobility, and sustainability; o Investing in research areas that further industry investments and support benefit opportunities; o Establishing Federal Motor Vehicle Safety Standards and infrastructure guidance as appropriate.

PRESENTATION	KEY HIGHLIGHTS

National Highway Traffic Safety Administration Address
 	
Speaker:
 	
Nat Beuse, Associate Administrator, Vehicle Safety Research, National Highway Traffic Safety Administration (NHTSA)
 	
Slides
	• NHTSA will determine how to ensure safety benefits are widely enjoyed and safety risks for crash avoidance technologies, including levels of automation are addressed.
	• The NHTSA levels of automation are intended to be a common framework for research or discussion, not a framework for regulation.
	• Significant NHTSA research activity exists for automation levels 0 and 1:
	  o Level 0: Radar, camera, and/or V2V enabled crash warnings
	  o Level 1: Radar, camera, potentially enhanced by V2V to provide level 0 capability plus single function automation
	• Currently conducting intensive research to support agency actions and evaluate technologies related to dynamic brake support, crash imminent braking, and pedestrian avoidance.
	• Key areas of NHTSA's automation research for levels 2 through 4 include: Human factors research, electronic control systems safety (including cybersecurity), system performance requirements, benefits assessment, and testing and evaluation.
	• NHTSA is releasing (August 18) an Advance Notice of Proposed Rulemaking and technical research report on V2V communications technology.

United States Department of Energy (Address Speaker: Patrick Davis, Director, Vehicle Technologies Office, Energy Efficiency and Renewable Energy, U.S. Department of Energy Slides	• Provided an overview of the U.S. DOE role and focus in vehicle research and development, and the synergies between automated and connected mobility and energy. • Foundational studies convey significant but uncertain energy implications of automation. • Possible future research includes expanding current activities to refine the foundational studies and provide a better and quantification of positive and negative energy outcomes. • Deeper research based on identified areas of opportunity may include transportation system modeling, the advanced vehicle design trajectory to 2050, and novel diagnostics, controls, and sensor development. • Business model innovation is as important as technology. • The U.S. DOE is not a regulatory agency in this space.

PRESENTATION	KEY HIGHLIGHTS
European Commission Address Speaker: Joakim Svensson Director , Volvo Group Trucks Technology Slides	• Provided overview of extensive recent, ongoing, and future research in automated vehicles and related topics, including communications and connectivity. • The Study SMART 2010/0064 project defined the necessary vehicle and infrastructure needed to support automated driving. • Automation projects under the Mobility for Growth call include AdaptIVe, iGame, Companion, and AutoNet 2030. • Under the new Horizon 2020 Programme, the next call for automation is a €20 million 2015 call for "Safe and Connected Automation in Road Transport." • International stakeholders have high interest in global standards to align approaches and synchronize efforts and avoid duplicating efforts.
Japan Ministry of Land, Infrastructure, Transport and Tourism Address **Speaker:** Takumi Yamamoto, Director, ITS Policy & Program Office, Road Bureau, Ministry of Land, Infrastructure and Transport **Slides**	• Japan's Cross-Ministerial Strategic Innovation Promotion Program includes a new research and development plan for an Automated Driving System. • The plan was initiated in May 2014 and includes 2.45 billion yen for fiscal year 2014. • Dramatic improvements are expected for the 2020 Tokyo Olympics. Targets include level 2 systems in 2017, level 3 in the early 2020s, and level 4 sometime thereafter. • There are three working groups which include both public and private members: system implementation, international cooperation, and next generation urban transportation.

California (CA) Department of Motor Vehicles Address
 	
Speaker:
 	
Bernard C. Soriano, Deputy Director, Risk Management, California Department of Motor Vehicles
 	
Slides
	• California Senate Bill 1298 mandates that the Department of Motor Vehicles (DMV) adopt regulations for Manufacturers' Testing and Operations on public roadways by January 2015.
	• The DMV's strategy includes two regulatory packages:
	  o Package 1: Submission of evidence of insurance, marking of vehicle on DMV database, and other feasible regulations (takes effect 9/14/2014).
	  o Package 2: Testing requirements, safety standards, operator license requirements, and vehicle registration requirements.
	• Regulatory Package 2 will not require a special driver or operator license.
	• Regulatory development efforts include significant outreach efforts: meetings with industry, presentations to government, public engagement through social media, traditional media, and working groups across jurisdictions.
	• It is important for NHTSA and state governments to work together to ensure consistency of state regulations before NHTSA develops Federal regulations.

Figure 10: Bernard Soriano, Deputy Director of Risk Management at the California Department of Motor Vehicles, discusses the state's efforts to regulate automated vehicle testing and operation (Image Source: Volpe Center).

Figure 11: Kevin Dopart, Program Manager for Connected Vehicle Safety and Automation at the U.S. DOT Intelligent Transportation Systems Joint Program Office, discusses U.S. DOT's efforts in connected and automated vehicle research during the U.S. DOT address at the 2014 Automated Vehicles Symposium (Image Source: Volpe Center).

Figure 12: Nat Beuse, Associate Administrator for Vehicle Safety Research at the National Highway Traffic Safety Administration, discusses the agency's research on automated vehicle safety during the 2014 Automated Vehicles Symposium (Image Source: Volpe Center).

5 Overview of Breakout Sessions

This section includes a brief summary of each breakout session's focus and goals, agenda, key issues and results, and next steps. These summaries were drawn as written from material provided by the organizers of each breakout session. The complete breakout session proceedings developed by the organizers of each group are provided as appendices. Table 1 below presents a list of the 10 breakout sessions at AVS 2014, as well as a link to the appendix with the full proceedings.

Table 1: List of Breakout Session Groups

	Breakout Session	Link to Appendix with Full Proceedings
1.	Evolutionary and Revolutionary Pathways to Automated Transit and Shared Mobility	Appendix A: Transit and Shared Mobility Proceedings
2.	Regional Planning and Modeling Implications of Driverless Cars	Full proceedings not available
3.	Roadway Management and Operations with Automated Vehicles	Appendix B: Roadway Management and Operations with Automated Vehicles
4.	Truck Automation Opportunities	Appendix C: Truck Automation Opportunities
5.	Legal Accelerators and Brakes	Appendix D: Legal Accelerators and Brakes
6.	The State and Future Direction of Automated-Vehicle Human Factors	Appendix E: The State and Future Direction of Automated-Vehicle Human Factors
7.	Near-Term Connected/Automated Technology Deployment Opportunities	Appendix F: Near-Term Connected/Automated Technology Deployment Opportunities Proceedings
8.	Personal Vehicle Automation Commercialization	Appendix G: Personal Vehicle Automation Commercialization
9.	Technology Roadmap, Maturity and Performance: Operational Requirements for Vehicle-Road Automation Systems and Components	Appendix H: Technology Roadmap, Maturity and Performance: Operational Requirements for Vehicle-Road Automation Systems and Components
10.	Road Infrastructure Needs of Connected-Automated Vehicles	Appendix I: Road Infrastructure Needs of Connected-Automated Vehicles

5.1 Evolutionary Pathways to Automated Transit and Shared Mobility

The Transit and Shared mobility breakout session examined and debated the evolutionary versus revolutionary path to automated transit and shared mobility. Invited speakers reported on current initiatives, vision, and policy. Discussion was structured to identify key issues along each pathway.

The evolutionary pathway discussion focused on current issues, near-term capabilities and 'low-hanging fruit', and possibilities for accelerating the evolution of automation to benefit existing transit services. Speakers addressed their vision for the role of automation within transit agencies, how specific forms of driverless technologies could be leveraged as they become commercially viable, and options for how agencies can prepare themselves (both internal organization and their interactions with external entities) so as to maximize the benefits as the technology matures. This *evolutionary* discussion focused on near-term opportunities for safety and collision avoidance as well as longer term visions and how to prepare.

The revolutionary pathway discussions addressed the critical path issues for major changes in public mobility such as systems of automated taxis, automated transit networks (ATNs), and shared mobility services. Speakers and participants focused on identifying critical developments and decision points to enable revolutionary change, critical milestones (technical, policy, or societal) and associated pivot points, possible bifurcation points, barriers, and possible solutions to identified barriers. The group concentrated primarily on an urban context, discussing the urban design implications for transitioning to automated mobility services.

Speakers

Day 1: Evolutionary Pathways

Alain Kornhauser, Princeton University
Sam Lott, Kimley-Horn and Associates, Inc.
Jerry Lutin, (retired) New Jersey (NJ) Institute of Technology and NJ Transit
Brian O'Looney, Torti Gallas and Partners
Louis Sanders, American Public Transportation Association (APTA)
Vincent Valdes, Federal Transit Administration (FTA) Office of Research, Demonstration, and Innovation

Day 2: Revolutionary Pathways

Adriano Alessandrini, CityMobil2, University of Rome La Sapienza
Ron Didiron, Sr., Mineta Transportation Institute
Neil Hoose, United Kingdom Transport Catapult
Joseph Kopser, RideScout and Chair, Defense Energy Center of Excellence
Christer Lindstrom, Institute of Sustainable Transport
Susan Shaheen, University of California Berkeley Transportation Sustainability Research Center

Summary of Discussion and Results

- AVs will further blur the lines between taxis, private vehicles and public transit. This trend is already in motion in the car and ride sharing industry, and with aggregators such as RideScout. As such the view of public transit should also evolve to that of providing public mobility, in the same vein of the Mobility on Demand (MOD) which emphasizes the person, and not the mode.
- Urban congestion is a complex problem. Automated single-occupancy vehicles (SOVs) alone are not the solution, and in many cases may exacerbate the problem. Methods to harness automated and driverless

technology for public mobility, in a form that contributes to a complete system (not just another niche mode), should be a priority moving forward.

- Applications of automation to existing modes, as well as realization of new modes are needed. Transit is a segmented market, with different solutions needed in different environments to create a fully integrated network. As such, innovators need to be cooperative, and not competitive in working with public authorities.
- Savings in liability costs as a result of collision avoidance technology on buses is forecasted to pay for the investment in as little as a year.
- Transit is a good, if not the best, environment to introduce automation; 2014 marks the 50th anniversary of the first fully automated transit system. Whether automated buses to improve line-haul, collision avoidance to minimize liability costs, or fully automated taxis to augment paratransit, the cost-benefit of centrally managed transit service can be assessed over the system lifecycle, and improvements quickly integrated to the improvement of the entire public mobility system.
- The first mile/last mile challenges are the opportunities for AVs, in the form of level 4 driverless operations. Existing line-haul system, whether high-speed trains, metros, or regional bus lines, can reasonably accommodate significant increases in demand, while our highways cannot. Efficient feeders and distributors to these systems (in various forms) have more potential to impact urban congestion in the near and far term than does other realizations of level 4 technology.
- This session advocates for the research of automated road vehicle technology to harness automation to architect livable spaces. For concepts such as transit-oriented development (TOD), complete streets or livable communities, vehicles are only tools to architect such spaces.

Next Steps

The breakout session identified key steps to keep the discussion of vehicle automation within public mobility as a key aspect of future transportation.

- Continue to identify and expose key issues related to the application of vehicle automation to public mobility by hosting periodic forums in cooperation with industry and in coordination with major events.
- Create a coalition to promote the key issues identified above to both the automate road vehicle innovators and to existing public transit. Members of the coalition will be drawn from attendees, and expanded as need and as key leaders are identified.
- Create an active partnership with Federal Highway Administration (FHWA), NHTSA, and other agencies involved in the Connected Vehicle Pilot Deployment Program to pilot application of vehicle automation to public mobility.
- Create and maintain a list of critical issues and initiatives, the origin of which are listed in key results above. This list of critical issues and initiatives will serve as the foundation for action and discussion within the coalition and at the forums moving forward.
- Advocate for specific advancements that will leverage automated vehicle technology for public mobility. At present this includes:
    - Research into the use of collision avoidance technology to reduce bus collisions, and subsequent liability cost as proposed by the Princeton Advanced Vehicle Engineering Research Center (PAVE).
    - Support the continuance of the memorandum of cooperation between Sweden and the U.S. DOT for continued research on ATNs.
- As public transit was identified as the most fertile area for implementation of automation and driverless technology, encourage analysis and planning studies of the potential impact of such technologies within the urban context of public mobility and congestion relief.

See **Appendix A: Transit and Shared Mobility Proceedings** for a full summary of this session.

5.2 Regional Planning and Modeling Implications of Driverless Cars

This session focused on the potential impact of driverless cars on our community and regional design/planning, built infrastructure development, transportation policies, and, at a more fundamental level, human activity-travel decision making. How may our activity participations and activity-travel patterns change, and how may city designs and land-use planning elements change to respond to changes in individual activity choices?

Unlike other workshops that have focused on technology aspects of automated cars and/or privacy and security considerations (which are all critical elements of a driverless car system and do have indirect impacts on behavior), the emphasis in this workshop will be on the mobility choices of individuals, households, and firms as the end-consumers of new technology on the one hand, and the decisions of car manufacturers, transit and transportation planning agencies, taxi systems, and freight carriers as system providers and intermediary-consumers of new technology on the other. The motivation for this workshop stems from the rather sparse attention on activity-travel behavioral impacts, and long-range regional transportation planning and policy implications, of driverless car technology.

The goals of the deliberations at the Symposium were two-fold: (a) To develop a limited set of plausible scenarios of autonomous vehicles in society; each scenario representing some combination of the dimensions of time frame, technology, lifestyle choices (such as all households owning no cars and purchasing services versus owning large autonomous vehicles), market groups of relevance (individuals and households, freight carriers, taxicab companies, etc.), and regulatory policies and public/private cooperative models (Day 1 Proceedings), and (b) To use the framing of the scenarios as a starting point to develop a modeling framework for analysis and planning (Day 2 Proceedings).

Speakers

Thomas Adler, Resource Systems Group, Inc.
Billy Charlton, Puget Sound Regional Council
Charlie Howard, Director of Planning, Puget Sound Regional Council
Jane Hayse, Director of Center for Livable Communities, Atlanta Regional Commission
Hani Mahmassani, Northwestern University
R. Jayakrishnan, University of California Irvine
Greg Larson, Caltrans
Eric Miller, University of Toronto
Srinivas Peeta, Purdue University
Steve Polzin, University of South Florida
Kevin See, Lux Research
Dale Thompson, U.S. DOT ITS JPO

Summary of Discussion and Results

- Developing specific scenarios is challenging at this stage because there are too many unknowns related to: Technology deployment, market adoption, lifestyle choices, business models, affordability, and changing generational travel behavior.
- No detailed data exist on traveler acceptance and understanding of technology – there are too many unknowns, and we haven't spent enough time thinking deeply about it.
- Automation level 3 – we have a pretty good handle on, we just need to think and work through appropriate analysis.
- Automation level 4 – this is where we need the most development work – both on understanding travel behavior changes, and how to model it.

- Working closer with policy makers will be important to understand their questions.
- Modeling issues to be resolved: car sharing effects on travel behavior; changes in household location; impacts to transit, walk and bike; etc.
- Increase in VMT seems likely – what are the policy implications for air quality, energy use, congestion, and land use.

Next Steps

- Organize a scenario development and vetting process to identify various pathways to market adoption and their effects of travel behavior.
- Conduct a synthesis of existing research on capacity and value of time implications of automated vehicles.
- Conduct special surveys, focus groups, and interactive gaming experiments to better understand human adoption and use of these technologies.
- A coordinated set of model runs in different geographic and modal contexts to understand the range of the magnitude of impacts.

5.3 Roadway Management and Operations with Automated Vehicles

Automated vehicles–at different levels of automation—will change traffic flow operations and management. As automated vehicles will drive differently than human-driven vehicles, several parameters such as time gap and reaction times will differ, affecting traffic flow capacity and stability. With roadside-to-automated-vehicle communication that enables transmission of signal timings and speed limits directly to vehicles, traffic management strategies will change. This breakout session aimed to identify the issues, opportunities and challenges for roadway management and operations with automated vehicles at near, mid and long term.

Speakers

Corey Clothier, Comet Robotics
Nick Cohn, TomTom
James Dreisbach-Towle, San Diego Association of Government
Edward Fok, Federal Highway Administration
Ram Kandarpa, Booz Allen Hamilton
Jamie Lian, Pacific Northwest National Laboratory
Shannon McDonald, Southern Illinois University Carbondale
Markos Papageorgiou, Dynamic Systems & Simulation Laboratory, Technical University of Crete
Bart Van Arem, Delft University of Technology, Transport Institute
Ismail Zohdy, Booz Allen Hamilton

Summary of Discussion and Results

- Automated vehicles are expected to increase traffic flow efficiency, but connected automation is required.
- Consideration of the vehicle as a sensor and traffic control actuator. How should they be regulated? What laws need to exist?
- Connectivity can bring early benefits, automation can leverage them.
- The traffic management context will change. These changes will include: travel/logistic patterns, type of vehicles, parking, and empty cars.

- The transition period is important
  - Mixed traffic, manual-automated transitions, managed lanes
- New management strategies needed
  - Centralized, decentralized, hierarchical, transactional
- Driverless vehicles will require new management approaches
  - Intersection control, platooning
  - Management of empty vehicles, Parking
- Dedicated lanes may accelerate deployment of driverless vehicles
  - Tracks in pavement, equity issues
- Lack of suitable modeling and simulation tools to accommodate automated vehicles (all levels)
  - Data, calibration, control models, information flows, driver behavior
- Traffic flow dynamics will fundamentally change
- Adapt from other disciplines (agriculture, power)

Next Steps

- The organizers will prepare a summary for follow up at the 2015 TRB Annual Meeting.

See **Appendix C: Roadway Management and Operations with Automated Vehicles** for a full summary of this breakout session.

5.4 Truck Automation Opportunities

The Truck Automation Opportunities breakout session focused on the discussion concerning how automated vehicle technology could affect commercial vehicle operation. It is important to note that commercial vehicles represent the transport of goods and services (e.g., truckload operations, less-than-truckload) and are covered by strict regulations. Myriad policies and regulations for commercial vehicle operation could be impacted by more advance levels of automation that are being introduced to the market. Hence, it becomes inevitable for the industry to consider and plan for possible upcoming changes while the technology moves along its maturation trajectory.

Speakers

Osman Altan, Federal Highway Administration
Fred Andersky, Bendix
RJ Cervantes, Director of Legislative Affairs, California Trucking Association
Deborah Freund, Federal Motor Carrier Safety Administration
Jeff Gonder, National Renewable Energy Laboratory
Mohammad Poorsartep, Texas A&M Transportation Institute
Joakim Svensson, Director of Volvo Group Trucks Technology
Dan Williams, Chief Engineer for Advanced Engineering, TRW Commercial Steering Systems

Summary of Discussion and Results

- Discussion took place regarding whether we should wait for V2V market penetration to catch up in order to enable platooning, or is it possible to deploy platooning application without V2V for two trucks driving in tandem. No agreement was reached in this discussion which raises the question if there are other alternative to Dedicated Short Range Communication (DSRC) based V2V to enable two trucks platooning.
- Key questions regarding implementation of CACC truck platoons that need to be answered are:

  - Driver - system HMI a key factor for user acceptance
  - Operating conditions
  - Dependability & liability
  - Mixed Fleets
  - How to form the platoon?
  - Can it be stepwise deployed?
- A need exists for studies focusing on following gaps. These studies include:
  - What following gaps are comfortable for drivers? The gap that may provide the best fuel efficiency may cause discomfort for the driver.
  - What the minimum required gap is to ensure the safety of the platoon, given the latency of sensors and V2V.
- A number of open questions remain:
  - What do truck fleets specifically need or want in terms of automation (e.g., dedicated infrastructure for level 3 automation and higher)?
  - How are owner/operator needs and challenges different from fleets in regards to these technologies?
  - What steps are necessary to alleviate other barriers to technology introduction such as legal, societal, or user perception?
  - What is the business case for each level of automation?
  - What are the upfront, ongoing, or hidden costs of commercial vehicle automation?
  - What is the incentive/disincentive? For example, are there insurance savings, decreased labor costs, or regulatory benefits?
  - What are the risks of early adoption?
  - What is the return on investment (ROI) where the owner of the truck is also the driver?
  - What infrastructure is needed for truck automation?
- Political considerations exist (e.g., Will automation eliminate good paying blue-collar jobs?)

Next Steps

Next steps identified for this breakout group and truck automation research include:

- Demonstrating the technology works and is safe and reliable
- Conducting pilots with customers, to verify potential market for CACC and potential fuel savings
- Creating standards
- Identifying and eliminate barriers (authorities, legal, market)

See **Appendix C: Truck Automation Opportunities** for a full summary of this breakout session.

5.5 Legal Accelerators and Brakes

This breakout session explored the legal issues that might accelerate or inhibit the deployment of autonomous and connected vehicles. The first two panels focused on different transportation modes, and the third discussed the regulatory environment in which these modes must function. Throughout the two days, participants were asked to focus on core questions relating to: (1) what mode is most likely to lead deployment in the current regulatory environment; (2) what role will "infotainment" play in driving or delaying deployment; (3) whether uniform laws would help or hinder deployment; (4) what opportunities and threats exist from increasing data use, and misuse; and (5) how can liability questions be best handled. The final part of the session was a group discussion that sought to prioritize the most important questions and provide recommendations for moving forward.

Speakers

Panel on Infrastructure—Tollways, Highways, and Transit:

Coco Briseno, Caltrans
Yeganeh Mashayekh, University of Pennsylvania
Francine Steelman, Miami-Dade Expressway Authority
Jeffrey Spencer, Federal Transit Administration

Panel on Vehicles—Military, Commercial, Truck, and Passenger:

Alan Korn, Meritor Wabco
Joakim Svensson, Volvo Truck
Ali Maliki, Ricardo
Barbara Wendling, Volkswagen Group of America

Panel on Government:

Jesse Chang, NHTSA
Maxime Flament, ERTICO
Timothy Mattson, Bowman & Brooke
Matthew Schwall, Exponent
Bernard Soriano, Deputy Director, CA DMV

Summary of Discussion and Results

The final portion session was spent reviewing key messages from the first three panels and translating that into key findings and recommendations for next steps

- Regarding the "lead mode" question, the group recognized the attention platooning has received, and several factors that may lead to early deployment of this commercial vehicle technology, such as potential cost savings and a cost-conscious industry that will push for its deployment. However, in thinking about what is likely to be the first higher-level deployment, low speed passenger shuttles appeared most likely.
- The group had some difficulty interpreting the role of infotainment. While several technologies appeared useful, and could develop attractive revenue streams, the group found as many potential legal problems in the current environment (e.g., distracted driving).
- Data and liability continued to be key areas of concern. While some idea of what data are currently protected, and what data should be protected in the future, is becoming clearer, there was a keen awareness that a significant data breach could significantly set back deployment.
- Similarly, liability appears to be acting as a potential brake on deployment, with concerns about lack of clarity in terms of the ultimate responsibilities of Original Equipment Manufacturers (OEMs) vs. after-market manufacturers, and lack of consensus regarding true "fail-safe" options (including skepticism of humans serving in that role).
- However, there was strong support for the creation of uniform laws, regulations or standards that would set a common "floor" for these technologies, with the belief that such a standard would accelerate deployment by creating a level playing field of common basic requirements and encouraging competition to exceed these standards at lowest cost. This enthusiasm, however, was tempered by the recognition that overly strict regulations could stifle further development and deployment.

Next Steps

- Given the advantages seen for developing a common regulatory "floor," steps toward considering and creating such measures should be undertaken.
- The lack of agreement regarding the potential liability risks and protections indicates a need for further dialogue among the legal and technology communities towards developing a common understanding about how risk and uncertainty in this area can be managed most effectively.

See **Appendix D: Legal Accelerators and Brakes** for a full summary of this breakout session.

5.6 The State and Future Direction of Automated-Vehicle Human Factors

The success of automated vehicles ultimately hinges on how well they meet their users' needs. The study and application of human factors throughout the automated-vehicle design cycle can yield a safe, useful, and reliable technology that does what its users want. The goal of this breakout session was to present the state of automated-vehicle human factors research and how it is being applied in the development of automated vehicles. The breakout session featured presentations from scientists and engineers leading the investigation of automated-vehicle human factors. Discussion panels provided attendees with a broad perspective of the issues and how they relate to one another. Attendees walked away with a better understanding of recent developments and what the immediate human factors research gaps are.

Speakers

Steve Casner, NASA Ames Research Center
Janet Creaser, University of Minnesota
Lutz Eckstein, RWTH Aachen Institute for Motor Vehicles
Greg Fitch, Virginia Tech Transportation Institute
Jim Foley, Toyota
Tsutomu Iwase, Subaru
Wendy Ju, Stanford University Center for Design Research
Johann Kelsch, German Aerospace Center
Michael Manser, Texas Transportation Institute
Michael Tsirhart, Visteon

Summary of Discussion and Results

- Operator interaction with partial automation is documented within aviation research. However, there are many significant differences between aviation and surface transportation, notably:
    - Pilots are highly-trained, and hand-selected, to operate a plane, where light vehicle drivers are relatively untrained and present more variability in capabilities and driving skills.
- There is substantial space between planes in the sky leaving pilots numerous seconds to respond to unforeseen events, where light vehicles are much closer to each other leaving drivers milliseconds to respond to unforeseen events.
- Driver misuse (unintentional improper use) and abuse (intentional improper use) of commercially-available partial vehicle automation is becoming a growing concern. There have been numerous videos posted online of drivers circumventing precautions vehicle designers have taken to ensure drivers remain in the loop with the driving task. Examples include drivers fastening soda cans to the steering wheel to trick the automation

into sensing that a hand is on the steering wheel. Research is needed to investigate real-world driver behavior with partial automation in order to characterize the different ways drivers misuse and abuse the technology.

- The effective and safe transfer of control from automated vehicles to drivers remains a top research priority. Simulator research has been performed since the previous workshop to investigate the human factors issues, but remains to be fully understood. The alert modality, timing of the alert, progression of alerts, and driver state need to be considered.

Next Steps

The next steps recommended by this breakout group include the following:

- Conduct a naturalistic driving study of driver use of commercially-available partial vehicle automation.
- Investigate the timing and modality of take-over-requests as well as the role of the environment and driver state in effectively regaining control of the vehicle when the automation becomes unable to operate in an upcoming scenario.
- Develop a comprehensive understanding of drivers' mental models of automated vehicle operation and what information is needed from the interface to inform/update these mental models.
- Investigate the role of driver monitoring to develop an equal understanding of the drivers' state as the state of the environment.
- Involvement of human factors research early on, as well as throughout, the design of automated vehicles to ensure user interaction with the automation is well understood and supported.
- Provide an update at the 2015 TRB Annual Meeting workshop.

See **Appendix E: The State and Future Direction of Automated-Vehicle Human Factors** for a full summary of this breakout session.

5.7 Near-Term Connected/Automated Technology Deployment Opportunities

Automated vehicles by themselves can deliver many but not all of the benefits possible through automation. Automation assisted by connection to other vehicles and the infrastructure and by roadway features such as managed lanes may be needed to unlock the full potential inherent in automation. Connected automation systems that include a role for an engaged driver can provide many benefits in the nearer-term, as part of the decades-long transition to a fully automated highway system. This breakout session will focus on ways to advance the development of an automated and connected highway system that will support increasingly capable vehicles and deliver safety and highway system operational benefits both in advance of and in addition to the benefits associated with fully automated vehicles.

It is important to identify these near-term opportunities in order to engage state and local highway transportation authorities and the automotive community. Public highway authorities for the most part tend to view fully automated vehicles as a consumer product that will deliver relatively few benefits for highway operations. Can incremental technologies that deliver near-term highway capacity and operational benefits help turn highway authorities from bystanders to advocates for an automated and connected highway system? If so, agencies can create a supportive technical and political environment for, among other things, the licensing and deployment of automated vehicles developed by the automotive community.

This session focused on near-term deployment opportunities and concluded with a discussion of how to integrate these near-term opportunities with ongoing efforts to deliver a fully-automated highway system.

Speakers

Technology Panel:

Kevin Dopart, U.S. DOT Intelligent Transportation Systems Joint Program Office
Robert Ferlis, Federal Highway Administration
Greg Larson, Caltrans
Michael Maile, Mercedes-Benz Research and Development North America
Christian Schumacher, Continental
Steven Shladover, California Partners for Advanced Transportation Technology (PATH)

Innovators Panel:

Colin Castle, Michigan Department of Transportation
James Dreisbach-Towle, San Diego Association of Governments
Casey Emoto, Valley Transportation Authority
Bob Frey, Tampa-Hillsborough Expressway Authority
Chuck Fuhs, Consultant
Ginger Goodin, Texas A&M Transportation Institute

Early Deployment Panel:

Thomas Bamonte, North Texas Tollway Authority
Doug Davenport, Prospect Silicon Valley
Steve Lockwood, Parsons Brinkerhoff
Suzanne Murtha, Atkins
Chris Schreiner, Strategy Analytics

Summary of Discussion and Results

Technology

- **Researcher perspective.** Actions needed to realize benefits: 1) Connected vehicle market penetration and 2) Focus on near-term level 1 applications – CACC, Speed Harmonization, and Eco-Signal Operation.
- **OEM and supplier perspective on connectivity and automation.** Vehicle connectivity is not going to enable automated driving but is going to make it better.
- **Deployment challenges from U.S. DOT perspective.** Aftermarket devices, spectrum sharing demands, communications congestion potential, other road users, and security.
- **Operational considerations from roadway authority perspective (State Department of Transportation).** Determine optimal balance of intelligence between vehicles and infrastructure; emphasize heavy vehicle applications as early adopters; and determine whether to start with mixed traffic or exclusive right-of-way.

Innovators

- **Agency Involvement.** Agencies – big and small – should be looking at this technology now to determine what the impacts will be. Agencies also need to be prepared to add capacity – staff, funding, technical capabilities.
- **Public role with private sector innovators.** One role is providing access to open operational data and leading education to build trust in technology.

- Identify public sector champions who need to address problems that automation can solve: Addressing congestion, providing additional capacity, and improving environmental impact.

Early Deployment

- **Roadway authority perspective.** Highway authorities can advance system technologies via investment/legislation/policy.
- **Identify value propositions for automated vehicles.** Safety and efficiency payoffs, policy supported by public benefit rationale, and increase in infrastructure capabilities.
- **Role of public sector.** 1) Establish facilities where we can learn more about the benefits of automation, and 2) sponsor and invest in early deployments.
- Be aware of consumer opinion and sensitivity to autonomous vehicles.
- Currently, consumers are wary of autonomous features but may not be fully informed of the potential benefits; willingness to pay is low and there are control/trust issues. However, safety can sell.
- Level 5 is the current consumer level of understanding and expectation; however, mobility benefits are achieved at level 1 and further levels of automation primarily offer consumer convenience – need to bridge this gap.

Transition to Long-Term Deployment

- Promising near-term (next ten years) deployment opportunities:
    - Level 1 applications for managed lanes and tollways
    - Public Transportation and Freight applications
    - Small scale community-based environments (retirement enclave, campus) offer opportunity for higher level automation deployment (level 4 and 5) due to lower risks
- **Changes in technologies, markets, economy, or society that must be considered.** Infrastructure investment, electrification, shared use, vehicle ownership sociology, liability and insurance framework for automation, international standards or guidelines, security / privacy / information sharing, stock versus aftermarket retrofit.
- How can we best transition from early deployment to long-term deployment?
- Accommodate progression of technology but allow for consistency with regard to performance – can standards minimize disruption and preserve benefits?
- Allow for incremental improvement– e.g., following distances for CACC.

Next Steps

The next steps identified by this group included the following:

- Circulate results to the relevant committees for development of research needs
- Encourage a structured dialogue including those committees responsible for developing research road maps
- Consider a task force to facilitate buy-in or identify roadway authorities that are currently participating or willing to participate in automation activities

See **Appendix F: Near-Term Connected/Automated Technology Deployment Opportunities Proceedings** for a full summary of this breakout session.

5.8 Personal Vehicle Automation Commercialization

Simply stated, the focus of the Personal Vehicle Automation Commercialization (PVAC) breakout session was to investigate how automated vehicles will make money.

The session addressed the commercial viability of automation SAE levels 3, 4 and 5 on passenger cars sold by the automotive manufacturers to consumers. Value elements of efficiency, safety and other factors were considered. Particular interest was devoted to new market forces such the insurance industry's positive or negative disruption from automation and the role of information, services and Infotainment companies who now have a new valuable channel to the consumer.

The goal of the PVAC session was and analysis of the various market forces that may impact the emergence of automated personal vehicles, including some that might not be intuitive in today's automotive industry.

Speakers

Chris Borroni-Bird, Qualcomm
John Capp, General Motors Director of Automotive Safety Office
Alex Mitchell, Automotive Industry Director, World Economic Forum
Michael Scrudato, Munich Reinsurance America
Kevin See, Lux Research
Kyle Vogt, Cruise Automation

Summary of Discussion and Results

- While the popular press has perhaps over-hyped the promise of vehicle automation in terms of how soon it will occur and how completely automated vehicles will be, significant progress is being made and the roadmap for launch of incremental automation features is well underway.
- The markets of shared vehicles and vehicle automation appear to be tightly connected, automation enabling viable business cases for shared vehicles and shared vehicle demand incentivizing the development and launch of automated vehicles.
- There was a strong consensus that SAE level 3 Conditional Automation, where the driver is expected to respond appropriately to a request to intervene, may not have a solid business case because it is unrealistic to depend on such behavior; however, SAE level 3 is a necessary step on the way to levels 4 and 5.
- Legislation and regulation were identified as challenges but workable over time.
- Myriad changes to society will occur with the advent of widespread vehicle automation.

Next Steps

- There is significant value in anticipating disruptive market launches due to perhaps currently unknown value propositions as well as anticipating barriers that may unexpectedly deter adoption. Further work would be valuable in investigating these factors in more detail, for example:
  - While the traditional auto manufacturers are proceeding with incremental feature launches, some members of the industry are advocating more aggressive, disruptive product launches. If a particular new automated vehicle offering resonates with a highly attractive business proposition, there could be acceleration of the market, at least for that particular vehicle configuration and application. What are some candidate business propositions that could create disruptive product launches?

- On the other hand, if market impedances dominate, such as legislative or regulatory issues, then even incremental launches of increasingly more highly automated vehicles could be delayed. What are come candidate market impedances that could delay AV launch?

See **Appendix G: Personal Vehicle Automation Commercialization** for a full summary of this breakout session.

5.9 Technology Roadmap, Maturity and Performance: Operational Requirements for Vehicle-Road Automation Systems and Components

Visions and concepts for automated vehicle-highway systems have as a common element the expectation of transformative personal and/or societal benefits. The performance of these future systems is fundamentally based on robust, reliable, safe and efficient technologies.

This session sought to examine state-of-the-art developments and technologies that will lend the needed support to realize the benefits of vehicle-road automation. Through presentations and group discussions, the session explored the needed performance, reliability and resiliency of these technologies using the framework of a broad and representative set of visions or states of automated vehicle-highway systems and their intended benefits.

Speakers

John Absmeier, Delphi
Roger Berg, Denso
Ching-Yao Chan, University of California Berkeley PATH
Jack Pokrzywa, Society of Automotive Engineers

Summary of Discussion and Results

These results capture the technology challenges based on group discussion of three future states of automation: (1) levels 3 and 4 automation, (2) mandated platooning, and (3) required connected capability.

- Levels 3 and 4 Automation:
  - Core technology challenges: Sensing and localization (HD maps and dynamic update); artificial intelligence; development of training software; integration with HMI; obstacle detection and avoidance in unanticipated scenarios (e.g., pedestrian on the highway); and passive and active safety designs for low-speed collision and coordination between systems).
  - Operational environment and policy challenges: Standard service scenarios (or driver-by driver with training) and standard obstacles and reaction times the manufacturer is not responsible for (vehicle will minimize harm).
  - Safety/reliability challenges: Skill decay and emergency/complex driving situations and license endorsement and retraining (simulator required?); development and testing of artificial intelligence software; complexity of system integration; data collection for validation; and redundancy systems.
- Platooning:
  - Core technology: Standardized networking; robust control (sensor noise/attacks/dropped packets); redundant systems (of same and different kind); and whole perception systems (especially urban perception) available for mass market.
  - Operational environment and policy challenges: Decision about whether to have continuous train or distinct platoons would be needed; sensitivity of sensing and communications to weather; willingness to pay for automation that is not available 100% of the time; developing joining and merging protocols; selection of an autonomous vs. human lead; and spacing to disband or transition.

- Safety/reliability challenges: Which failure modes can be accommodated; tradeoff between safety and close spacing; adaptive maneuvering regimes for heterogeneous s vehicles; and regulations to standardize when automation can be used.

- Required Connected Capability:
  - Core technology: Bandwidth and cognitive radio and notification (without display?)
  - Operational environment and policy challenges: Use of infrastructure or p2p; channel sensing to alert user of non-operational status or overload; defining functionality before radio technology, with consideration of what is being communicated; and consideration of how non-users can benefit.
  - Safety/reliability challenges: Understanding the effect of frequency on use.
- Cross-Cutting Safety and Reliability Challenges:
  - Whether there is Federal, state, OEM, and third-party testing is a regional preference that varies across countries.
  - How to deal with aging and re-certification remains an open question.
  - Model certification (aerospace level of testing/performance would be cost prohibitive). Testing and creating data on all possible scenarios would not be possible.
  - An adversarial environment must be considered in the certification process: outline standard threats.
  - Identified three degraded modes: fail safe, fail operational, and fail soft.

Next Steps

- Complete More In-Depth "for the public good" use case approach
- Technology requirements for different operational policies
- Examine available versus necessary technologies
- Consider standards, certification, and regulation
- Fill gaps with research, investment and policy including: safety, robustness and reliability needs; innovation needed; and necessary policy steps.

See **Appendix H: Technology Roadmap, Maturity and Performance: Operational Requirements for Vehicle-Road Automation Systems and Components** for a full summary of this breakout session.

5.10 Road Infrastructure Needs of Connected-Automated Vehicles

This breakout session discussed how digital and physical infrastructure can support connected-automated vehicle technology. Day 1 focused on digital mapping of the infrastructure, including data needs, collection, availability, accuracy, and reliability of data. Day 2 was directed towards physical infrastructure advancements, such as enhanced roadway markings and intelligent traffic signals, in an environment where connected-automated vehicles are present.

The goal of the session was to better understand infrastructure and related data needs for connected and automated vehicles from different stakeholder and industry perspectives. We worked to identify opportunities and challenges in both digital and physical infrastructure as well as stimulate discussion about research needs, opportunities for innovation, and potential next steps to prepare infrastructure to be supportive and compatible with automated vehicles. The session focused on the following key questions:

- What is included in the definition of digital infrastructure and how does it support connected-automated vehicles? Is it more than just mapping? What infrastructure related data are needed and how are they obtained? What are the hurdles faced in collecting and maintaining these data?
- How can asset management practices today be integrated with data needs for connected-automated vehicles? What changes to existing practices would be required in order to support these vehicles?

- How can physical infrastructure advancements support connected-automated vehicles? How can vehicles use both physical and digital infrastructure in an integrated fashion?
- What technical and policy challenges exist when discussing digital and physical infrastructure on a national scale for connected-automated vehicles? What research is needed to address these challenges?

Speakers

Day 1—Digital Infrastructure:

Robert Dingess, President, Geospatial Transportation Mapping Association
Alexander Klotz, Vice President Business and Technology Innovation, Continental
Serge Lambermont, Technical Director of Automated Driving, Delphi
Ogi Redzic, Vice President of Connected Driving, Nokia HERE
Ron Singh, Oregon Department of Transportation

Day 2—Physical Infrastructure:

Kevin Balke, Research Engineer, Texas A&M Transportation Institute
Paul Carlson, Division Head Traffic Operations and Roadway Safety, Texas A&M Transportation Institute
Ronald Gibbons, Director of the Center for Infrastructure Based Safety Systems, Virginia Tech Transportation Institute
Chris Harris, Civil Engineering Manager, Tennessee Department of Transportation
Ray Mandli, President, Mandli Communications
Jim Misener, Independent Consultant
Tony Tavares, Chief of Division of Maintenance, Caltrans

Summary of Discussion and Results

The following are key points taken from both days of discussion on physical and digital infrastructure.

- States are collecting digital data for various purposes, including asset management, performance evaluation and safety analysis. Industry is also collecting digital data to generate high definition maps for navigation and vehicle automation purposes. There may be an opportunity to "gather once, use often" and share data to support the deployment of automated vehicles and achieve benefits for both the public and private sectors.
- Additional discussion is needed between industry and states to identify what data and related requirements are necessary to support connected-automated vehicles in order to determine if leveraging State collected data are viable.
- Currently States are collecting data in using various collection standards and data formats; standardization of these processes may be required to share data or generate a common base map.
- States are responsible for deploying and maintaining traffic control devices. States are willing to make changes to infrastructure in support of automated vehicles if guidance is provided. This requires more discussion with industry and automated vehicle developed to determine requirements and priorities for physical infrastructure.
- Implementation of traffic control devices can vary state by state. Today's guidelines, such as the Manual on Uniform Traffic Control Devices (MUTCD), provide several alternatives for implementation; some of which may be more challenging than others for automated vehicles. Revision to standards and guidelines may be necessary.
- Traffic control devices are designed for human comprehension; multiple alternatives pose a greater challenge for automated vehicles.

- Low cost infrastructure changes could be implemented now; other changes can be factored into long term planning

Next Steps

The discussion of infrastructure needs seems to be only in the beginning stages. The following next steps were proposed in order to begin discussing what infrastructure changes may be required for both digital and physical infrastructure in order to facilitate the deployment of automated vehicles.

- Hold workshops at the national and state level to gather requirements from developers of automated vehicles
- Assess impacts of requirements on digital infrastructure data collection and maintenance of physical infrastructure
- Develop recommendations and guidance to share across all states to accelerate deployment of automated vehicles

See **Appendix I: Road Infrastructure Needs of Connected-Automated Vehicles** for a full summary of this breakout session.

6 Overview of Ancillary Sessions

6.1 U.S. DOT Listening Session

The U.S. DOT is carrying out research under a five-year research plan to address the federal research role in automation in road transportation. The goals of the U.S. DOT Multimodal Program Plan for Vehicle Automation are to:

- Identifying benefit opportunities in automated vehicle technology;
- Facilitating development and deployment of automated transportation systems that enhance safety, mobility, and sustainability;
- Investing in research areas that further industry investments and support benefit opportunities;
- Establishing Federal Motor Vehicle Safety Standards and infrastructure guidance as appropriate.

The U.S. DOT stakeholder engagement event at AVS 2014 provided an opportunity for U.S. DOT staff to present on active research projects related to vehicle automation and to and receive feedback from the diverse set of stakeholders present at the symposium.

PRESENTATION	KEY HIGHLIGHTS
Developing a 5-Year Program Plan for Vehicle Automation – Kevin Gay, Volpe Center	• Internal U.S. DOT and external stakeholder input served as a foundation for developing a 5-Year Program plan. • Subsequent activities translated this input into a draft Automation Program vision and goals, objectives, research needs, projects and milestones. • Next steps in the process include ITS and modal budget analysis, finalizing project scopes, and developing a program roadmap. • The program plan and roadmap for automation will align with the forthcoming U.S. DOT ITS Strategic Plan.
Accessible Transportation Technologies Research Initiative (ATTRI) – Mohammed Yousuf, FHWA	• In the United States there are 56 million people with disabilities-plus large and growing elderly and wounded warrior populations. • Automation could improve mobility choices and options, increase flexibility, and enhance independence of travelers with disabilities. • ATTRI identifies, coordinates, develops, and implements new integrated solutions in advancing such capabilities. • The project is completing Phase I: Exploratory and User Needs Research in September 2014. • Phase II: Innovation and Prototype will begin in September 2014. Phase III: Demonstration will run from 2017-2019.
Framework for Estimating the Benefits of Automation – Dale Thompson, ITS JPO	• The project goal is to develop a framework to estimate the potential safety, mobility, energy, and environmental benefits of automation on the United States surface transportation system. • Review and coordination with existing work in these areas is a foundational element of the project. • Metrics identification will include identification of target applications, metrics for measuring, and assumptions for the modeling framework.

	• The final product will propose a modeling framework including: gap analysis, recommended models, suggestions for integration, and a proposal for a prototype model development. • The framework will not be final in 12 months; it will be ready for review and input.
Improving Safety Through Automation, Program Update – Paul Rau, NHTSA	• The major goal is to improve motor vehicle safety by defining the requirements for automation assisted driving that is: o Functionally safe and electronically reliable; o Secure from malicious external control and tampering; o Precise in vehicle steering, braking, and acceleration; o Compatible with driver abilities and expectations. • A project focused on drivers transitioning into and out of automated driving states enabled by level 2 and level 3 automated driving concepts is nearing completion. • Two upcoming research projects are Functional Safety of Automated Lane Centering Controls and Target Crash Populations for Automated Vehicles.
New Research in Truck Platooning – Osman Altan, FHWA	• Described the three U.S. DOT funded truck platooning research projects. • The Partial Automation for Truck Platooning project examines technical and policy issues. Includes two, three-truck platoon demonstrations with Volvo trucks. • Heavy Truck CACC System – Testing and Stakeholder Engagement for Near-Term Deployment: Examines platoon operations, interoperability, human factors, fleet operations acceptance, and system robustness. Includes a two-truck platoon demonstration with Peterbilt trucks. • Naturalistic Study of Truck Following Behavior: Examines truck following behavior in relation to the development of automation technologies.

Question/Comment Topics

After each of the presentations, the U.S. DOT responded to questions and comments from attendees. The dialogue highlighted the diversity of interests among participants and provided useful input for the U.S. DOT's consideration. Topics of inquiry from attendees included the following:

- Cybersecurity for infrastructure
- As vehicles and infrastructure become more intelligent, there may be a need for funding for local intelligent infrastructure maintenance and operations
- Consideration of a transition phase in the 5-Year Program Plan for automation
- Consideration of the needs of mobility impaired travelers
- Consideration of potential regional variation in the benefits of automation and automation-related policies
- The role the ITS JPO role in standards for automation
- The U.S. DOT vision and view of level 3 automation
- Automated collision avoidance in trucks

- Inclusion of travel time reliability as a performance metric in the benefits framework project
- Clarification of the intended audience for the benefits framework project
- Consideration of safety disbenefits/unintended impacts in the benefits framework project
- Potential changes in the behavior of non-automated vehicle drivers
- Consideration of the macroeconomic impacts of automation
- The inclusion of local agencies in outreach efforts for U.S. DOT automation activities
- The unique considerations of automated vehicles at international border crossings
- The relationship between automation and Metropolitan Planning Organization (MPO) long-range planning requirements

6.2 Society of Automotive Engineers on Road Automated Vehicle Standards Committee

The open meeting on Automated Vehicle Standards and Best Practices provided an opportunity for attendees to learn more about the work of the On Road Automated Vehicle Committee of SAE International. This session was about standards and best practices for automated vehicles on public roads. Specifically, an overview of the work being undertaken by the SAE On Road Automated Vehicle Committee in this spirit was be presented, followed by a panel of experts discussing specific technical task groups and activities of the committees. Topics included definitions and taxonomy, safety testing, verification & validation, human factors, interoperability, and legal ramifications of on road automated vehicles.

The panel discussion was followed by an open forum for audience participation to discuss issues, standards, and best practices for automated vehicle. The session agenda was as follows:

1. Welcome, Committee Overview| Paul Perrone, Chairman, ORAVS Chairman
2. Panelist Discussion
    - Definitions and Taxonomy – Barbra Wending, Senior Principal Engineer, Volkswagen Group of America
    - Safety Testing – Steve Underwood, Director, Connected Vehicle Proving Center at University of Michigan – Dearborn
    - Verification and Validation—Paul Perrone, CEO, Perrone Robotics
    - Human Factors – Amy Klinkenberger, Senior Safety Regulatory Engineer, Hyundai-Kia America Technical Center
    - Reference AVS Architecture & Interface – Dan Bartz, Associate, Booz Allen Hamilton
3. Audience Discussion | Paul Perrone, moderator

6.3 Poster Sessions

The symposium included two poster sessions to highlight research in the area of road vehicle automation.

Session 1 featured 18 peer-reviewed posters. Posters included in this session present a discrete and completed research project, with clear findings and/or conclusions. Posters were judged by workshop participants. The winning poster was "Situational Awareness and Levels of Autonomous Driving" from Stanford University. The University of Leeds provided a $100 prize for the award. The table below lists the posters and provides a link to the poster if available.

Table 2: Poster Session 1 Summary

ID	Title	Authors	Organization(s)
1	Situational Awareness and Levels of Autonomous Driving	David Miller, Annabel Sun, Brian Mok, Mishel Johns, Tongda Zhang, Yeon Joo, Key Lee, David Sirkin, Wendy Ju, Clifford Nass	Stanford University
2	Agent-based Modeling and Simulation Framework for Automated Vehicles	Monstasir M. Abbas, Miloš N. Mladenovic	Virginia Tech Transportation Institute
3	Enhancing Active Safety by Autonomous Driving Intelligence System Based on Experienced Driver Behavior Model	Pongsathorn Raksincharoensak, Masao Nagai, Hideo Inoue, Minoru Kamata, Masayuki	Tokyo University Japan; Automobile Research Institute, Toyota Motor Corporation, University of Tokyo, Toyota Central Research &Development Lab
4	Travel and Environmental Implications of a Shared Automated Fleet, with Varying Levels of Automation	Daniel J. Fagnant, Kara Kockelman	University of Texas at Austin
5	Enforcing Liveness in Autonomous Traffic Management	Tsz-Chiu Au, Neda Shaidi, Peter Stone	Ulsan National Institute of Science and Technology; University of Texas Health Science Center at Houston, University of Texas at Austin
6	Quantitative Benefits and Costs of Vehicle Automation	Yeganeh Mashayekh, Chris Hendrickson, Jeremy Michalek, Constantine Samaras	Carnegie Mellon University
7	Autonomous Driving – Aspects of Acceptance in a Sociotechnical Transformation Process	Eva Fraedrich, Barbara Lenz	Humboldt University Berlin, German Aerospace Centre
8	Assessing the Energy Impact of Dedicated Lanes and Vehicle Platooning	Vadim Sokolov, Dominik Karbowski, Namwook Kim	Argonne National Laboratory
9	Energy Use Implications of Partial Light Duty Vehicle Automation	Yeganeh Mashayekh, Chris Hendrickson, Jeremy Michalek, Constantine Samaras	Carnegie Mellon University
10	The Tension Between Autonomous Cars' Impacts on Junction [Intersection] Capacity and Occupancy Comfort	Scott Le Vine, Alireza Zolfaghari, John Polak, Stan Young	Imperial College of London, University of Maryland

ID	Title	Authors	Organization(s)
11	Valuing the Convenience of Autonomous Vehicles	Daniel Morton, Cody Kamin	SingaporeMIT Alliance for Research and Technology
12	Vehicle Pre-clustering: An Intersection Control Strategy with Emerging Communication and Control Technologies	Xiaopeng Li, Fang Zhou, Jiaqi Ma, Mengqi Hu	Mississippi State University, University of Virginia
13	Assessing the Potential of Automated Vehicle Technologies in Expanding the Consumer Market for Battery Electric Vehicles	Jing Dong, Changzheng Liu, Zhenghong Lin	Iowa State University, Oak Ridge National Laboratory
14	Lawyers and Engineers Can Speak the Same Robot Language	Bryant Walker Smith	Center for Internet and Society at Stanford Law School, University of South Carolina Schools of Law and Engineering
15	A Review of Merging Strategies for Automated Vehicle Platoons	Lin Xiao, Raymond Hoogendoorn, Bart van Arem	Delft University of Technology
16	Modeling the Energy Use of a Connected and Automated Transportation System	Austin Brown, Jeffrey Gonder	National Renewable Energy Laboratory
17	Class 8 Tractor Trailer Platooning Effects on Fuel Economy	Mike Lammert, Jeffrey Gonder	National Renewable Energy Lab

Poster Session 2 included 22 posters presenting a wide range of topics and activities including introductions to research centers or programs, research agendas, prospective studies, and work in progress.

Table 3: Poster Session 2 Summary

ID	Title	Authors	Organization(s)
1	Developing Laboratory Test Capabilities to Supplement On-Road Autonomous Vehicle Development	Craig Pavlich, Eric Rask	Argonne National Laboratory
2	Vehicle Automation and the Law of the Newly Possible	Bryant Walker Smith	Center for Internet and Society at Stanford Law School, University of South Carolina Schools of Law and Engineering
3	Potential Applications of Currently Available Automated Transit Network Technologies	Reuben Juster, Robert Johnson, Stanley Young	University of Maryland, R.E. Johnson Consulting

ID	Title	Authors	Organization(s)
4	Autonomous Vehicle Technology: A Guide for Policymakers	James Anderson, Nidhi Kalra, Karlyn Stanley, Paul Sorensen, Constantine Samaras, Oluwatobi A. Oluwatola	RAND Corporation, Cambridge Systematics
5	Using Cooperative Adaptive Cruise Control to Form High-Performance Vehicle Streams	Steven E. Shladover, Xiao-Yun Lu, Christopher Nowakowski, Robert E. Ferlis	California PATH Program, University of California Berkeley, Federal Highway Administration, Turner-Fairbank Highway Research Center
6	Partial Automation for Truck Platooning	Xiao-Yun Lu, Steven E. Shladover, Christopher Nowakowski, Matt Hanson	California PATH Program, University of California Berkeley, California Department of Transportation
7	Human Factors of Automated Driving: Towards Predicting the Effects of Authority Transitions on Traffic Flow Efficiency	Silvia Francesca Varotto, Raymond Hoogendoorn, Bart van Arem	Delft University of Technology
8	Road Weather and Automated Vehicles	Sudharson Sundararajan, Ismail Zohdy	Booz Allen Hamilton
9	FHWA Office of Operations Research and Development Connected Automation Research	Joseph I. Peters, Anna Giragosian, Tom Phillips	Federal Highway Administration Office of Operations, Leidos
10	Southeast Michigan 2014 Connected Vehicle System Implementation	Walton Fehr, Greg Krueger, Frank Perry, James Marousek, Tom Lusco, Dave McNamara	U.S. DOT ITS JPO, Leidos, Booz Allen Hamilton, Iteris, MTS
11	HF Auto - Human Factors of Automated Driving : Legal Aspects and Market Perspective of Highly Automated Driving	Miltos Kyriakidis, Joost de Winter, Riender Happee	Delft University of Technology
12	The Role and Design of Communication System for Automated Driving	Gaurav Bansal, Yaser P. Fallah, Mohammed Fanaei, Amin Tahmasbi-Sarvestani, John Kenney, Matthew C. Valenti	Toyota InfoTechnology Center, West Virginia University
13	Demonstrating How to Reduce Oil Consumption and Achieve Emission-Free Mobility as Connected, Automated Vehicle Travel Rises	Steve Marshall, John Niles	Center for Advanced Transportation and Energy Solutions
14	Macro Economic Impact of Autonomous Vehicles	Richard Mudge	Compass Transportation and Technology
15	Using Driving Simulation to Examine Human Factors Issues in Vehicle Automation	Anuj K. Pradhan, C. Raymond Bingham, John Sullivan, Ryan Eustice	University of Michigan Transportation Research Institute

ID	Title	Authors	Organization(s)
16	Connected and Autonomous Vehicles 2040 Vision	Yeganeh Mashayekh, Allen Biehler, Chris Hendrickson	Carnegie Mellon University
17	Automated Driving in Relation to Today's Use and Ownership of Cars	Eva Fraeddrich, Barbra Lenz	Humboldt-University Berlin, German Aerospace Centre
18	Regulatory Issues and Potential Strategies for More Highly Automated Vehicles	Christopher Nowakowski, Steven E. Shladover, Ching-Yao Chan, Han-Shue Tan	California PATH Program, University of California Berkeley, California Department of Motor Vehicles
19	Crash-Imminent Scenarios and Driver Models for Safety Evaluation in Vehicular Automation	Arda Kurt, Ümit Özgüner	The Ohio State University
20	Towards a European Roadmap on Key Technologies for Automated Driving	Gereon Meyer	VDI|VDE Innovation + Technik GmbH / EPoSS Office
21	Research Activities on Automated Driving in Germany	Adrian Zlocki, Philipp Themann, Felix Fahrenkrog Devid Will, Lutz Eckstein	Institut für Kraftfahrzeuge (ika), RWTH Aachen University
22	Evaluation Methodology for Supervised Automated Driving Including Impact Analysis on Safety and Environment	Adrian Zlocki, Felix Fahrenkrog Lutz Eckstein	Institut für Kraftfahrzeuge (ika), RWTH Aachen University

6.4 Ancillary Workshop—Envisioning Automated Vehicles within the Built Environment: 2020, 2035, 2050

This ancillary all-day workshop presented the opportunity to focus on the policy and built environment issues related to automated vehicle use. The workshop committee focused the event on assisting MPOs in their efforts to address these new technologies. The workshop was financially supported by: University of California Davis, National Center for Sustainable Transportation, Southern California Association of Governments, ARUP, Kimley Horn, Fehr & Peers, and the National Center for Intermodal Transportation.

Steve Shladover started the day with an overview of AVS 2014, and then in order for the workshop attendees to have a base of knowledge several handouts were distributed titled Technology Application Pathways to Commercialization. In the interest of addressing the built environment three generic design sites were created: Streets and Roadway, Neighborhood and District and Regional for use in the hand-on workshop. Seven scenarios were developed and participants had the ability to choose the scenario group in which they wanted to take part. 110 participants took part in some or all of the day's agenda, representing a wide set of backgrounds and disciplines. Nine scenario groups formed each with a committee member and one or two facilitators to enable the discussions and drawings.

In the interactive workshop, participants discussed and envisioned the built environment impacted by automated vehicle technologies. Small, multi-disciplinary teams of experts combining knowledge of a wide range of fields - city planning, infrastructure and architecture, car design, engineering, software and systems – collaborated on six different chosen scenarios. Each team started with all three design sites (streets/roadway, neighborhood/district and regional scales) described in words, images, maps and diagrams. Scenarios were developed with a time-horizon and presence of specified types of vehicles with varying degrees of shared mobility, connected vehicle technology, and autonomous operation. The

Scenario groups generated, through different interactive methods and writing/visualizing techniques, "after" scenarios in order to think through the challenges and benefits to our built environment that an autonomous mobility future can hold. Each team briefly presented their outputs at the end of the workshop for feedback and discussion with the wider set of participants.

See **Appendix J: Friday Ancillary Workshop—Envisioning Automated Vehicles within the Built Environment: 2020, 2035, 2050** for a full summary of this breakout session.

7 Symposium Wrap-Up

At the conclusion of the workshop, the organizing committee met to consider next steps. The event was a success as measured by the size of the registration, and the level of sustained engagement of participants at the plenary, breakout, and ancillary sessions. In response to demand from the community, the organizers are planning to produce a fourth annual symposium in 2015.

Appendix A: Transit and Shared Mobility Proceedings

Session and Proceedings Organizers: Dan Fagnant (University of Utah), Reuben Juster (University of Maryland), Alain Kornhauser (Princeton University), Walt Kulyk (FTA), Scott Le Vine (Imperial College of London), Rachel Liu (New Jersey Institute of Technology), Shannon Sanders McDonald (Southern Illinois University Carbondale), Nazy Sobhi (U.S. DOT Volpe Center), Stanley E. Young (University of Maryland)

I. Session Focus and Goals

The Transit and Shared Mobility session examined and debated the evolutionary versus revolutionary path to automated transit and shared mobility. Invited speakers reported on current initiatives, vision, and policy. Discussion was structured to identify key issues along each pathway.

In the evolutionary pathway explored primarily on the first day, speakers and discussion focused on current issues, near term capabilities and 'low-hanging fruit', and possibilities for accelerating the evolution of automation to benefit existing transit services. Speakers addressed their vision for the role of automation within transit agencies, how specific forms of driverless technologies could be leveraged as they become commercially viable, and options for how agencies can prepare themselves (both internal organization and their interactions with external entities) so as to maximize the benefits as the technology matures. This evolutionary discussion focused on near-term opportunities for safety and collision avoidance as well as longer term visions and how to prepare.

In the second day, the discussions addressed the *revolutionary* pathway – the critical path issues for major changes in public mobility such as systems of Automated Taxis, Automated Transit Networks, and Shared Mobility Services. Speakers and participants focused on identifying critical developments and decision points to enable revolutionary change, critical milestones (technical, policy, or societal), and associated pivot points, possible bifurcation points, barriers, and possible solutions to identified barriers. The group concentrated primarily on an urban context, discussing the urban design implications for transitioning to automated mobility services.

II. Day 1 Presentations and Discussion

Opening Remarks

The presentations consist of representatives from a broad spectrum of public transit interests including the Federal Transit Administration, APTA, New Jersey Transit, the consulting community serving automated people movers and automated metros, and the development community. Although the transit and shared mobility crowd is a minority at the Automated Vehicles Symposium, it is an important minority, representing how automation can be harnessed for visions of mobility beyond personal automobiles.

Panel/Presenter Summaries

Speakers	Summary of Presentation, Question and Answer Session
Vincent Valdes, Associate Administrator Research, Federal Transit Administration	Mr. Valdes opened with how Intelligent Vehicle-Highway Systems has evolved into Intelligent Transportation Systems, reflecting additional modes of transportation beyond personal automobiles. Likewise, the initial focus on automation for the personal vehicle will give way to how to harness driverless automobiles for the mobility of the public. Mr. Valdes outlined a vision forming within the FTA's research group of a MOD concept. MOD takes advantage of changes in social media, demographics, population shifts, preferences, trends, and technology. MOD puts a premium on the person, being traveler centric rather than mode or vehicle centric. It begins with interoperability and data exchange. It will evolve into a multimodal decision support system, spontaneous mobility, and eventually an automated transportation system. Similar to a recent ad slogan for cellular networks, in which the pitch line is 'It's the network ....' Similarly the focus of transit needs to shift to realize that connecting people efficiently will require seamless transitions between modes and across technologies, requiring a shift in mentality of our transit management from mode specific to people centric. In response to a comment about the poor utilization of on-demand buses, Mr. Valdes commented that MOD (along with technology advances) has the potential to make buses as we know them obsolete unless they adapt, while heavy rail systems under MOD would provide the needed 'line-haul' link. The last mile problem appears to be the most urgent issues which technology is bested suited to address.
Dr. Jerry Lutin, Retired, New Jersey Transit & New Jersey Institute of Technology	Dr. Lutin began his presentation with a striking statistic, travel by buses is getting safer, but causality and liability costs are increasing rather than decreasing. It is estimated to cost $8,000 per bus per year. About 6% of the total operating cost is spent to address claims arising from accidents and collisions. This estimate is from publically submitted data from transit agencies, and is likely below the actual costs as several associated expenses are not included in those tabulations. Collision avoidance systems currently available as automotive packages could decrease these liability costs, and have the potential to pay for themselves (that is the cost of procurement and outfitting the bus fleet) within a year of operations. However, the bus fleet presents unique challenges. The lifespan of a public transit bus is between 12-18 years, far exceeding the practical lifespan and supportability of the automotive electronics. That long lifespan can be a problem since computer technology often becomes obsolete in 18 months to two years, making maintenance and parts replacement problematic. The technology solution to implement collision avoidance needs to have open architecture and standards so operators are not trapped into proprietary systems, and have feasible upgrade paths when equipment exceeds design life. Dr. Lutin shared an on-going research initiative that will demonstrate collision avoidance technology on bus fleets, which would include creating autonomous/collision avoidance bus technology requirements and standards so private industry can develop systems and public transit can leverage the cost competitiveness of open, standard based systems.

Speakers	Summary of Presentation, Question and Answer Session
Louis Sanders, Director of Technical Services, American Public Transit Association	Mr. Sanders opened his presentation with a reflection on APTA members' attitude toward driverless technology as applied to public transportation. "Some transit agencies cannot wait for automated technology to arrive, and some cannot wait for it to go away." Mr. Sanders reflected that public transit operators' prime operating hours are at peak period. Although buses and transit run throughout the day, the majority of riders, similar to vehicles, are in the peak hours. This occurs at a time of day when roads are at capacity and parking is limited. The low hanging fruit for transit is to have automated vehicles / systems serve the last mile of commuting trips, expanding the catchment basin for existing line-haul systems, many of which can be easily expanded with additional rolling stock to service greater customer base. Driverless technology for road based systems will likely not have an impact in the near term because, driverless or not, the road system is at capacity during peak hours. An audience member asked if transit agencies are working more together. Mr. Sanders answered with the example of how San Francisco Municipal railways merged into the San Francisco Municipal Transportation Agency, a prime motivator of which was to curb the mode specific management, and the beginning looking at providing trip based perspective to public transit.

Speakers	Summary of Presentation, Question and Answer Session
Sam Lott, Kimley-Horn and Associates, Inc.	Mr. Lott opened his presentation with a brief review of the development of the automated people mover industry, noting that 2014 marks the 50th anniversary of the introduction of the first fully automated prototype system in the transit industry. Mr. Lott focused on the possibilities for transit guideway operations as automated and connected roadway vehicles are integrated into guideway (or dedicated transitway) systems. As one of the examples of the maturing of the automated guideway transit industry, he discussed how the oldest heavy rail line in Paris – Régie Autonome des Transports Parisiens (RATP's) Metro Line 1 – was converted to fully automated, driverless operation in 2013. Further evidence of the Automated People Mover (APM)/Automated Transit System industry maturity are the several standards for automated guideway transit operation. The International Electrotechnical Commission (IEC) has adopted the Automated Urban Guided Transport Safety Requirements (IEC 62267) and the American Society of Civil Engineers (ASCE) APM (ASCE-21). The ASCE APM Standard has recently added off-line station provisions, which are quite relevant to the merging of automated roadway vehicle technology into guideway transit through an evolutionary process. Mr. Lott covered a robotic bus system Toyota piloted a decade ago in Toyota's demonstration project during the 2005 Aichi World Expo. Based on a published technical paper authored by Toyota engineers, these robotic buses could electronically couple and uncouple for dynamic platooning of the buses on the fly, and the automated buses could switch between manual and automated operations. Such automated transit vehicles would eliminate the need for costly rail switches (i.e., the robotic vehicle would steer themselves and would not need physical guidance or switches). Mr. Lott's comments and observations were also influenced by a technical paper he coauthored for the 2005 APM Conference. The premise for the paper was heavily influenced by a Group Rapid Transit (GRT) operational analysis he performed for Bay Area Rapid Transit (BART), and the paper was coauthored by the head of BART's Research and Development Department. These two technical sources that documented projects involving dynamic transit routing and dispatching shaped his conclusions on the application of automated, robotic buses (i.e., automated/connected roadway vehicles) for line haul use. If vehicles/trains were dynamically coupled and decoupled as they progressed through a transitway system, and if off-line stations were created to facilitate demand responsive service, the operational costs and origin-to-destination travel times would significantly decrease, while the passenger level-of-service would significantly increase. Transit agencies could run a dedicated transitway in a hybrid manner with partially fixed routes and partially demand-responsive flexible service. He closed his presentation by advocating that the transit industry provides the best environment to test automated roadway vehicle operations, and many in the audience affirmed his insight.

Speakers	Summary of Presentation, Question and Answer Session
Brian O'Looney, Design Architect/ Partner, Torti Gallas and Partners	Mr. O'Looney opened with a film clip of San Francisco's Market Street before the 1906 earthquake, highlighting how in a pre-auto dominated realm, people lived the streets very differently than today. Automation offers a similar transformation of how we live in our street realms. After World War 2, Americans grew to love cars and suburbs, and car manufacturers capitalized on this with advertising that associated vehicle ownership with freedom and fun. In current times, a measureable shift in car ownership attitudes is occurring, particularly in the young and old. The younger generation and retirees, particularly those living in dense urban areas, see cars more as a burden (with ownership costs and challenges) rather than a symbol of freedom. Automation technologies provide an opportunity for cities to return the streets to pedestrians. Mr. O'Looney warned the audience that while automation would result in a considerable amount of good effects, there are many other concerns and risks from the externalities of full-automation of private vehicles. Circulating automated recreational vehicles could further clog our street system with people looking for a new form of affordable housing, where fuel to continually cruise streets is cheaper than rent. Automated vehicles could serve as advertising – constantly circulating highly populated areas similar to planes trailing banners observable on our beaches. Such unforeseen uses of driverless technology, along with increased traffic from "deadheading" vehicles and media/technology fed "vehicular swarming" (Big 2 hour sale at Macys or Madonna and Bono are at the corner of 5th and 42nd-Let's Go!), would result in increased grid lock. Since automated vehicles will be programed to avoid collisions, pedestrians need not to follow traffic laws, and could choose to jaywalk and therefore initiating automated vehicles' emergency stop sequence, perhaps leading to increased fencing and barriers in the urban realm, like that now proposed for College Park, Maryland, to prevent such behavior. To resolve all of this, our urban cores are likely to experience the implementation of intensive variable tolling for private vehicles and a robust automated multi-modal public transit system.
	Discussion primers – Prior to open discussion, Dr. Jerry Lutin and Dr. Alain Kornhauser prepared and shared 'Perspectives on the Elephant in the Room' – that is the potential of automated and driverless transit to create major paradigm shifts in concepts of public mobility.

Speakers	Summary of Presentation, Question and Answer Session
Dr. Jerry Lutin, Retired, New Jersey Transit & New Jersey Institute of Technology	From his multi-decade experience researching, teaching and practicing transit planning, Dr. Lutin shared his view on automation's disruptive, yet accelerative force to Transit Oriented Development and similar concepts. Transportation demand modeling has evolved with external pressures for change. Some of the first transportation models included the 1960s era FHWA highway network model focused on how vehicles move through systems. The 1970s yielded the Urban Mass Transit Administration's Transit Planning System and addition of the logit mode split model to reflect transit ridership. The 1980s resulted in alternatives analysis within the planning process. With the passage of the Intermodal Surface Transportation Efficiency Act in the 1990s, MPOs were empowered and localities had to model air quality impacts, particularly in non-attainment areas. TOD, parking and VMT reductions, and trip chaining were new focuses. During the past decade, complete streets, non-vehicular travel, healthy communities, and activity based models were new trends that impacted the planning process. In 2020 and beyond, automation will be the new focus. Dr. Lutin concluded his discussion with a view how the classic four step model would be impacted by different levels of the NHTSA levels of automation.
Dr. Alain Kornhauser, Professor, Princeton University	Dr. Kornhauser's commented on evolutionary and revolutionary changes to transit enabled by automated technology, emphasizing the commercialization paths and reinforcing that transit is one of the best markets for level 3 and 4 technologies. Current active collision avoidance could save transit agency large sums in liability, and in turn acclimate transit to automation. Automated vehicles fit well in retirement communities where high speeds are not needed, and ability to efficiently serve short trips is already demonstrated in the proliferation of golf carts in such communities. Automated vehicle operation could begin operation on less complex local streets where most people start their trips (at their homes). The New York City initiative to convert 42nd street into low-floor light rail is another possibility that low-speed vehicle automation could address. Large investments in infrastructure are not required to begin to leverage automated vehicle benefits. One to two vehicles could be used to feed line haul transit in New Jersey, capitalizing on New Jersey Transit's excess rail capacity. The consumer market for NHTSA level 4 is speculative, because no one will want a car that they cannot drive. The first experience with level 4 automation may very well be transit, or a vehicle co-op.

Summary of Breakout Session Discussion

- What should the industry do as a whole to work together to implement low hanging fruit enabled by automation and driverless technology?
    - Industry needs to promote this type of conversation, not just a single breakout session yearly. Within a consortium, transit needs to identify objectives, enabling technologist to provide meaningful products.
- Members of the Community Transportation Association of America (CTAA), which has more members than APTA, are most likely to be first adopters of automated technology with respect to para-transit service. They need to be brought into the consortium.

- Recognize that the transit industry as a whole is a soft spoken industry, and not skilled in communicating to the public. They are adept at communicating with their customers, but not at communicating with potential customers. An illustration is transit agencies' trip planners which have a great tendency to malfunction, while Google Transit works more predictably. In moving forward with driverless and automated technology, Transit should seek partners skilled for this role.
- The standards process in transit lags much of industry as exemplified by Bus Intelligent Transportation Systems that do not talk to each other. APTA and FTA, combined with other industry players, need to identify and support an appropriate standards process and consortium.
- Integration of transit information should proceed equal in priority to implementations of automation and driverless technologies. The United States could learn from Europe where SuperHub (Europe) brings together transit services similar to Mobility on Demand. It aggregates information from all types of mobility (transit, bike share, car share) including cost and timing information. It is implemented in 9 different European countries and transportable to other geography.
    - Note that Transport Direct's (United Kingdom equivalent to SuperHub) biggest challenge was keeping transit timetables up to date.
    - Capitals TTS (pre-SuperHub) ended right away since it was too expensive and funding died, illustrating the challenge for such integration.
- Paratransit was identified as a low-hanging fruit for the first application of driverless technology. However, for paratransit, many times the driver's function also includes assisting passengers getting in and out of vehicle. Technology has yet to provide an easily accessible automobile (or van or bus) for the disabled.
- What role do transit companies have in a totally revolutionized world?

Summary/Highlights of Day 1 Presentations and Discussions

- Mobility on Demand, a concept for public transit that puts a premium on the person, is enabled by automated and driverless transport.
- A near-term, low-hanging application is crash avoidance for public buses– completely paid for by reduced liability.
- APTA – 'We have members that can't wait for it to happen, and members that can't wait for it to go away.'
- First mile and last mile presents strategic opportunities where AVs can directly impact public transit, typically as line-haul systems have excess capacity that can be brought on line in response to increased ridership.
- Automation in transit recently celebrated its 50th anniversary ... lessons learned are transferrable (standards, safety, design, and human factors) to automated vehicles.
- Automated buses with electronic coupling eliminate switching in line-haul transit, opens a world of possibilities for increased capacity and efficiency.
- AV can return the street to the people IF appropriately designed.
- Transit is a good (if not the best) environment to introduce automated road vehicle technology.

III. Day 2 Presentations and Discussion

Opening Remarks

- Dr. Dan Fagnant opened day two with a brief summary of the previous day presentations and discussions, and introducing the revolutionary pathway. This pathway refers to both the introduction of new means of public conveyance as well as the associated business models that depart from the traditional government owned, maintained, and operated methodologies of providing mobility apart from single occupancy vehicles. The speaker lineup for day two consisted of a variety of perspectives from initiatives currently in progress that provide a glimpse of revolutionary change that is either occurring or possible.

Speakers	Summary of Presentation, Question and Answer Session
Dr. Adriano Alessandrini, Researcher, University of Rome La Sapienza	Dr. Alessandrini shared experiences from the CityMobil2 project emphasizing that current automation capabilities have application in urban areas now, provided that the appropriate planning and enabling legislation and regulation is in place. Shared mobility will have its main benefits only from full automation (removing the driver completely will allow different transit and shared mobility concepts that are impossible today). Full automation on every street has still a long way to go and requires critical choices and potential liability problems. It is however possible to automate the last mile as done today with CityMobil2, if vehicle automation is done with consideration of (and sometime adaptation to) the environment. The CityMobil2 consortium derived from the rail technical standards a risk assessment approach and adapted to automated road transportation systems to insert them safely (and hopefully legally) in urban environments. It consisted of multiple steps including, the identification of the possible hazards and the necessary mitigation measures to the engineering of vehicles, infrastructure and control systems, mitigating the hazards, testing vehicle and system fail safe reactions, and testing (and limiting when necessary) the operational conditions. Dr. Alessandrini emphasized the need for United States contemporaries that are advocating for public mobility to not ignore this process, and to initiate now the public discussion needed to create enabling legislation. Overall vehicle automation will need a much more a collective effort.
Joseph Kopser, CEO and Co-Founder, RideScout	Mr. Kopser shared his perspective as a leading figure in transportation aggregators, that is websites and information services that provides easy access to any transportation option available to the public, be it public transit (buses and rails) or private offerings like Zipcars and other shared mobility services. This is an emerging business field in which companies are vying for market share similar to the travel aggregators that have emerged in the last decade for hotels, airlines and car rental. RideScout, and like companies, provide a similar service for local and regional transportation, reducing the complexity for consumers to recreate local or regional trip without a car, or a traditional car rental. Important takeaways from Mr. Kopser's presentation are that this trend is fed by the demographic shift from the younger generation away from car ownership, is not reliant on captive ridership like traditional transit offerings, and these are commercial ventures, not subsidized. This trend is happening now without automated or driverless functions. It is fueled by the mobile data revolution that connect people to services through their smart phones, enabling real-time connection of mobility offerings on the fly, not just pre-planned trips arranged primarily from desktop oriented web site. Automated transit

	offerings, as well as driverless vehicles do not enable RideScout, but rather will become mobility alternatives that will fit into existing mobility aggregators' frameworks, and is anticipated to enhance the market already enabled by smart phone technology and existing market demand for non- vehicle-owned mobility options.
Dr. Susan Shaheen, Adjunct Professor, University of California Berkeley	Dr. Shaheen provided a perspective of the current car and ride sharing phenomenon, and how it will be impacted by automation and driverless technology. Similar, to Mr. Kopser's perspective, car-sharing is a current market initiative enabled both by information and communication technology as well as growing demographic shifts away from car ownership. Car sharing is for short-term vehicle access, unlike car rentals characterized by a minimum of a single day use. People are trading off car ownership for shared access, typically available 24/7. Examples of this type of service include Zipcar and Car2Go. Another form of car-sharing are on demand ride-sharing (sometimes termed transportation network companies) where everyday drivers respond to other people's request for transportation, aligning both the drivers and passenger request so that the route accommodates both. Although the destination implies ride-sharing to a common location, or locations along a common route, the ride-sharing industry has emerged as a disruptive model to traditional taxis service, generating many lawsuits. Examples of this type of ride-sharing company include Uber and Lyft. Similar to the RideScout presentation, the car and ride-sharing businesses that have emerged are not enabled by vehicle automation, rather they exist independent of vehicle automation. However, automated and driverless vehicles has the potentially accelerate both of these emerging business models. Level 3 capability has the potential to greatly reduce the liability costs involved in such services, further lower the entry to market capital and operating costs. Level 4 capability introduces the ability to re-position vehicles to respond to demand, and even provided automated-taxi like service in the long term. Level 4 capability will further blur the lines between ride-sharing and commercial taxi service, though not expected to be a viable market force in the short term. Dr. Shaheen shared an overview of her Integrated Active Transportation System integration study which included several different alternative "worlds" to plan and how automated technology influenced each alternative.
Rod Diridon, Sr., Executive Director, Mineta Transportation Institute	Mr. Diridon exhorted the breakout session that in pursuing the new technology (which we should) we cannot disregard the programs and infrastructure that currently exists and is being implemented. The new concepts enabled by technological advances are needed and must mature (such as driverless vehicle technology), and must be integrated to complement existing systems. The population in California is projected to double within the planning horizon. Without a system of transit oriented developed centered on a sustainable rail backbone AND a system of transportation to link the first and last mile enabled by automation and driverless transit, the vision for future of California is not bright. With growing carbon dioxide emissions, whose primary contributor is surface transportation, California cannot address climate change while still relying on petroleum powered

	vehicles (shared or un-shared). Silicon Valley has 65 train stations, enough space to build 50 story skyscrapers to prevent urban sprawl, and could accommodate double the population – if we can solve the first-mile / last-mile problem with automated vehicle technology in concert with non-petroleum based power. Lastly, Mr. Diridon, as an experienced politician and public leader admonished the group to advocate together, not in competition. The demand is larger than the supply. It is much more difficult for elected officials to turn down initiatives when presented a united front. If technologists appear in competition (rather than cooperative), denial of initiative is much more easily accomplished.
Prof. Neil Hoose, Head of European Engagement, Transport Systems Catapult	Dr. Hoose reported on an initiative to solve the first/last mile problem at Milton Keynes, a planned city about 45 miles northwest of London. The town is served by rail on the south. The project objective is to design and demonstrate a distributor system to connect the train to end destinations, as well as interlink the activity centers. Dr. Hoose shared concepts explored including automated vehicles that traverse pedestrian pathways. He set out some of the challenges in terms of sensing and intelligent actions and the notion of the "character" of the vehicle. Discussion evoked debate on why the project targeted using pedestrian pathways as opposed to roads, which in turn discussed the issue of returning common areas to pedestrians. The Milton Keynes project is exemplary of the first/last mile issue to which current transit technology provide no efficient solution, and for which automated and driverless technology may first address.
Christer Lindstrom, Founder, Institute of Sustainable Transportation	Mr. Lindstrom shared perspectives on automated transit networks with respect to the needs of Stockholm, Sweden. Stockholm currently serves 70% of commuter needs with transit and is one of the fastest growing areas in Europe. Despite its high transit ridership, auto ridership is growing during off peak periods, however the topology of Stockholm consisting of interlinked islands does not allow for increased road construction. This has led to a philosophy in planning that is summed up by the slogan – 'no more rubber on the roadway', acknowledging that growth in automobile usage is in conflict with the growth and health of the city. Automated transit networks, a form of automated, driverless vehicles running on dedicated transit ways is being investigated heavily as a means to meet public mobility without compromising health of the cities. ATNs can cut carbon emission as well as petroleum dependence (strategically important for Sweden with respect to energy vulnerability and the current political situation in Russia). Studies indicate ATNs are slightly more expensive than bus, but with significantly more benefit. The self-driving car has merit for Stockholm, but with limited road infrastructure resources, it provides no long term strategic solution. A Memorandum of Cooperation between the United States and Sweden is currently being enacted for continued ATN research and implementation.

Summary of Breakout Session Discussion

- The emergence of mobility aggregators such as Ride Scout, as well as the emergence of ride-sharing and car-sharing in concert with smart phone proliferation, and each proceeding forward without the need for automated concepts speaks to the cost of information complexity in our public mobility scene. These concepts rely not on new modes, but on using technology to reduce either the information complexity or cost model complexity to the average consumer – allowing more consumers to participate. Will automated vehicles compliment this movement even further, allowing transit to become a true utility, fading into the background of our society's fabric much like our power, water and sewer systems today?
- The ride and car sharing models appear to have carved a new sector that is neither public nor private, or perhaps blurred the lines between the two. Will automation further blur these lines, and possibly provide greater opportunity for private investors to serve the public mobility market, something that is currently unattractive (or more appropriately – unprofitable) in our current public transit system.
    - People will pay a premium for high quality service, and this provides an opportunity for private capital. However, current models of public transit are void of opportunities for private participation, and as such one of the emphasis areas should be to prepare for private investment in public mobility systems, both from a policy and public support perspective.
    - Government's primary role should be seen to set minimum standards, and articulate those standards to the market, letting the private marker respond as automation comes on line. For example, to proliferate car-sharing, public can promote with 'car-sharing only' parking spaces. Likewise, more 'pay-per-use' policies (tolls, fees, etc.) would better reflect the true cost of SOVs to the public, encouraging more consideration (and thus larger market) for public mobility options.
- RideScout and its competitors bring choice and competition to the public transportation information space, monetizing (with private capital) public transit information dispersal. Each company differentiates itself with distinctive styles and use patterns, and in so doing, optimizes the availability of data to the public. Non-participation by Uber and Lyft, similar to Southwest within the airline industry, is not a major hindrance.
- Private entrance into public mobility is already happening with buses. This includes brand new private bus companies that place a strong emphasis on real time information. There are 27 transit operators in the San Francisco Bay area. Private operators for big companies (such as Google) place in the top 10 with respect to size of fleet.
- Insurance for level 4 car sharing is a key issue to be addressed.

Summary of Day 2 Results/Session wrap-up

- Lessons learned from CityMobil2: the policy and legal framework is essential in the process to leverage automated vehicle technology in the public mobility arena.
- Shared vehicles, shared rides, and seamless integration of choices across platforms (as enabled by information aggregators such as RideScout) are growing elements in our fabric of non-SOV choices enabled by wireless information technology. These elements will include and leverage AVs when available, but will continue to grow without AVs.
- Attitudes toward car ownership are changing measurably, as auto-ownership is viewed as a burden and less as an icon of freedom.
- Externalities (global warming, threat of terrorism, and energy dependency) may significantly shape AV applications.

- European cities, as exhibited by Stockholm, are seriously investigating AV to enable an entirely different system of automated mobility (ATNs), motivated by constrained land-use environment, concerns over energy dependency and its political ramifications, and long-term environmental sustainability.
- First-mile and last-mile connectivity repeatedly arose as the application ripe for innovation with AV, and would provide the greatest overall benefit in synergism with existing transit line-haul infrastructure, leveraging our current and planned infrastructure along with new technology.
- The demand for public mobility far exceeds supply, leaving room for multiple, integrated solutions. Advocacy in partnership, as opposed to competition, will be the most effective in the public arena.

IV. Key Results

- AVs will further blur the lines between taxis, private vehicles and public transit. This trend is already in motion in the car and ride sharing industry, and with aggregators such as ride scout. As such the view of public transit should also evolve to that of providing public mobility, in the same vein of the Mobility on Demand which emphasizes the person, and not the mode.
- Urban congestion is a complex problem. Automated SOVs alone are not the solution, and in many cases may exacerbate the problem. Methods to harness automated and driverless technology for public mobility, in a form that contributes to a complete system (not just another niche mode), should be a priority moving forward.
- The demand for public mobility exceeds current supply. Applications of automation to existing modes, as well as realization of new modes are needed. Transit is a segmented market, with different solutions needed in different environments, to create a fully integrated network. As such innovators need to be cooperative, and not competitive in working with public authorities.
- Safety enabled by automation is as much a motivator as driverless operations. Savings in liability costs as a result of collision avoidance technology on buses is forecasted to pay for the investment in as little as a year. Such commercialization avenues are the focus of the newly announced Princeton Research Center at Monmouth.
- Transit is a good, if not the best, environment to introduce automation; 2014 marks the 50th anniversary of the first fully automated transit system. Whether automated buses to improve line-haul, collision avoidance to minimize liability costs, or fully automated taxis to augment paratransit, the cost-benefit of centrally managed transit service can be assessed over the system lifecycle, and improvements quickly integrated to the improvement of the entire public mobility system.
- The first mile / last mile are the opportunities for AVs, in the form of level 4 driverless operations. Existing line-haul system, whether high-speed trains, metros, or regional bus lines, can reasonably accommodate significant increases in demand, while our highways cannot. Efficient feeders and distributors to these systems (in various forms) have more potential to impact urban congestion in the near and far term than does other realizations of level 4 technology.
- Lastly, the transit and shared mobility session advocates for the research of automated road vehicle technology to harness automation to architect livable spaces. Concepts such as TOD, complete streets or livable communities, the vehicle as well as the automated vehicles are only tools to architect such spaces.

V. Next Steps

The breakout session identified key steps to keep the discussion of vehicle automation within public mobility as a key aspect of future transportation.

- Continue to identify and expose key issues related to the application of vehicle automation to public mobility by hosting periodic forums in cooperation with industry and in coordination with major events. Industry participants include FTA, ATPA, TRB, American Planning Association (APA), ASCE, Advanced Transit Association (ATRA), CTAA, Association of Metropolitan Planning Association (AMPO), AUVSI and others. Major events include association annual meetings (APTA, CTAA, AMPO, & APA).
- Create a coalition to promote the key issues identified above to both the automated road vehicle innovators and to existing public transit. Members of the coalition will be drawn from attendees, and expanded as need and as key leaders are identified.
- Create an active partnership with FHWA, NHTSA, and other agencies involved in the Connected Vehicle Pilot Deployment Program to pilot application of vehicle automation to public mobility.
- Create and maintain a list of critical issues and initiative, the origin of which are listed in key results above. This list of critical issues and initiatives will serve as the foundation for action and discussion within the coalition and at the forums moving forward.
- Advocate for specific advancements that will leverage automated vehicle technology for public mobility. At present this includes:
  - Research into the use of collision avoidance technology to reduce bus collisions, and subsequent liability cost as proposed by the Princeton Research Center.
  - Support the continuance of the memorandum of cooperation between Sweden and the United States Department of Transportation for continued research in Automated Transit Networks as shared by Christer Lindstrom.
- As public transit was identified that the most fertile area for implementation of automation and driverless technology, encourage analysis and planning studies of the potential impact of such technologies within the urban context of public mobility and congestion relief.

Appendix B: Roadway Management and Operations with Automated Vehicles Proceedings

Session Organizers: Bart Van Arem (Delft University of Technology), Ram Kandarpa (Booz Allen Hamilton), Edward Fok (FHWA), George Munteau (Pacific Northwest National Laboratory), Kevin Heaslip (Utah State University), Siva Narla (Institute of Transportation Engineers), Nick Cohn (TomTom), Markos Papageorgiou (Technical University of Crete), Maxime Flament (ERTICO), Robert Bertini (Portland State University), Sudharson Sundararajan (Booz Allen Hamilton), Ismail Zohdy (Booz Allen Hamilton)

I. Session Focus and Goals

Automated vehicles – at different levels of automation- will change traffic flow operations and management. As automated vehicles will drive differently from human driven vehicles, several parameters such as time gap and reaction times will differ, affecting traffic flow capacity and stability. With roadside to automated vehicle communication that enables transmission of signal timings and speed limits directly to vehicles, traffic management strategies will change. This breakout session aimed to identify the issues, opportunities and challenges for roadway management and operations with automated vehicles at near, mid and long term.

II. Day 1 Presentations and Discussion

Opening Remarks

The focus of this session was on traffic operations and management when there are automated vehicles in which drivers still have major control (NHTSA level 3 and below). Audience will be asked to provide expected changes to the freeway and arterial/urban networks in the next five to ten years and expected impact of penetration of level 3 automated vehicles on roadway traffic management and operations. We had 4 speakers representing government, private and research community to set the stage and prepare participants for panel presentations and sub-breakout discussions that will follow the panel.

Panel/Presenter Summaries

Speakers	Summary of Presentation, Question and Answer Session
Ram Kandarpa Senior Associate Booz Allen Hamilton	Mr. Kandarpa introduced topics related to both freeway management and arterial management. The focus of the presentation was to invoke questions to understand how automated vehicles will affect operations related to various aspects of freeway and arterial management. For example how will different levels of automation affect traffic operations? How should Transportation Management Centers train personnel to accommodate and handle different levels of automation?
Prof. Markos Papageorgiou Director Dynamic Systems & Simulation Laboratory Technical University of Crete	Dr. Papageorgiou spoke about how centralized and decentralized traffic control will help various levels of automation. The presentation was filled with examples and scenarios that outlined the advantages and disadvantages of a centralized versus Decentralized approach.
Nick Conn Tom Tom	Mr. Conn presented how roles of different traffic management entities will change and adapt as different levels of automation penetrate the market place. He particularly distinguished between road operators and service providers and what their respective goals and demands will be. For example service providers will focus more on parking services, eco-routing, weather forecasts for road

	travel etc. while government agencies will focus on safety, optimal use of infrastructure and cost reductions.
James Dreisbach-Towle San Diego Association of Government	Mr. Dreisbach focused on automated vehicles and integrated corridor management (ICM). He highlighted the benefits one can expect from ICM and what technology adoption means to transportation agencies. He provided an anticipated timeline for various levels of automation to penetrate the market place and introduced the idea of in-vehicle active routing.

Summary of Panel Discussion

Discussion Topic 1 "Freeway Traffic Management and Operations"

- Physical/Functional Architecture
  - How should traffic management be structured? How the functional architecture work, especially if there is no current design available.
  - Decision making is supposedly a real-time process and this should be the anticipated future of applications
- The decision making process is critical and it is going to be distributed among cars using certain methods.
  - The decision can be in a vehicle-basis or more generalized for the traffic flow level.
  - There is a reason to have a decision making on-board even if it includes travel management advisory/regulation. There is a big move to go towards hybrid system for on-board capabilities.
- If the traffic management center (TMC) wants to assess a situation or incident, it is important in the troubleshooting process to accommodate the decision making to know what happened.
- It is important to consider archiving and identifying the individual vehicle performance to monitor the traffic violation.
- Can we impose a specific speed limit to the autonomous vehicles? Wide range of possibilities we have and we need to mention that when we study the regulations.
- Recurrent/Non-recurrent Conditions
  - Now the recurrent/non-recurrent specification is not an issue any more, it is a matter of traffic management. We already took care of the recurrent conditions and we need to deal with non-recurrent.
  - In fully autonomous environment, we can easily optimize the system using connectivity.
  - Rubbernecking, multi-tasking increases the congestion significantly. For example, Mercedes talk today mentioned that putting driver aside eliminates the impact of human distraction and enhances the flow/throughput
- We can think more from the management perspective, definitely there are many things we can do to enhance the system.
  - Definitely, incident management in autonomous environment is much efficient and faster to recover from any incident/crash.
  - How to deal with mixed environment? Having non-automated vehicles along with automated ones is an issue.
  - Definitely, the breakdown of automated vehicles will be a new type of incidents that we might have in the future
- Road Users
  - The Applications for the Environment: Real-time Information Synthesis (AERIS) program, the environmental enhancement and fuel consumption is definitely a motivation for the equipped vehicles. We found some good results even with a small level of penetration

- In terms of Risk, we need to compare the perception of driving a car where you have control versus not.
- Aggressive drivers are comfortable to drive high-speed cars. So young generations have different opinion and they are more comfortable with high-end cars.

- In the urban planning, especially at some cities, you cannot get rid of your car and you need it to transport. Maybe in the next 20 years, everything will change and everything will be remotely available
  - The question now, who is the owner of the car? My insurance can reduce the money that I am paying because they can monitor my driving. So in the future, we need to think what will happen if the owner is not in control anymore.
  - What is better: automated vs non-automated? I would say based on the task or the application. So there is no specific correct answer.
- The best management objectives: Safety or efficiency?
  - It is not option, they are not mutually exclusive. We need to consider both.
  - We might improve safety but we might create new type of safety issues that were not exist
  - I did not say that "either/or" option. We can have both together. The main intention that we can aim enhancing both options with different levels.
- Behavior change over time is a question mark. There may be an unintended increase in VMT.
  - I would use the airline example; enhancing airline system should be similar to enhancing vehicle system in terms of assuring safety for all drivers.
  - To sum-up, a distinguished speaker from Australia mentioned we evaluate deaths using cost, and we evaluate similarly for efficiency.

<u>Control Actions for Freeways</u>

- Pricing – probably a major controlling factor
  - Implementation issues
  - Affects majority of the population
  - Unified pricing, travel systems – ultimately beneficial to all travelers (outcome from Helsinki)
  - Does automation provide unified pricing?
- Agent based models
  - Space time reservation systems
  - Priorities for certain vehicles
  - Efficiency increases
- How do we change the way we study freeway operations?
- Subject topic is very fundamental to freeway operations
  - Ability to control specific vehicles
- Significantly increased level of control with increased automation → privacy issues?
  - How do you deal with the public?
  - Success – worldwide adoption? But different countries are very different
  - Could possibly be abused?
- Digital signatures are important for identification and control – currently missing

<u>Discussion Topic 2 "Urban and Arterial Traffic Management and Operations"</u>

<u>Signalized intersections</u>

- Enhancements to arterial operations?
  - Level of performance depends on the nature and characteristics of the application

- More applications in arterial environment might offer more benefits
- Gains with very small steps (2-5%)

- Distinguish between V2I and V2V when we talk about automation?
- Automated vehicles optimizes signals as they travel the corridor
- Route guidance and route choice?
  - Some drivers try to avoid left turn movements (even protected)
  - Can impact some turning movements
  - Benefits depend highly on assumptions being made (arrival / departure rates?)
- Any minimum penetration rate that will provide operational benefits?
  - Signal-less intersection – all vehicles need to be equipped

Transit and freight operations

- Spread transit out but you want to platoon freight
- Wide spectrum to deal with in case of freight – different loads
- Automation will help freight safety more
  - Lot of applications already exists – Automated drive train (fuel efficiency)
  - Don't have to automate every aspect of driving but need to pay more attention to routing
  - Improve freight efficiency
  - Road side inspections can become more efficient
- Transit and freight are normally larger vehicles – important for traffic safety view point
- Automated vehicles(transit and freight) to enhance data quality for TMCs
  - Possibly transmit computed data as well – use computation capability of the vehicles
- Reliability of freight systems as increase in number of goods to be delivered
- Need for large transit and freight vehicles might go away
  - Deliveries at night
  - Inefficiencies could go away?
- Dynamic permitting – for route changes
- Change for transit – might be a big challenge > cost prohibitive > making it not a very good business initiative
  - Not from central transit authorities – financial issues

Parking management

- Smart phone parking
  - Park your car with your smart car – gives you more space
  - Change the way we design our cities or infrastructure – could be transformational
- SF Park (in San Francisco)– Parking demand management
  - Remote garages etc. → last mile problem?
    - Car drops you off and parks itself
  - Can a vehicle make its own parking reservation?
- Parking policies?
  - Can we just make the car go for a drive for an hour instead of paying $25 or more for 1 hour parking? Cheaper? → What does this to the environment?
  - Physical configuration could change

  - Push back from the valet industry
  - Push back from government entity because of loss of revenue?
- Travel behavior changes?
  - Empty cars trying to get into the city to pick people up to go home all at the same time thereby creating a bottleneck?
- Using energy from electric vehicles at parking structure?
  - Automation with wireless charging – renewable sources for electricity
- Concepts of car sharing – eliminates need for parking?
- Who provides parking information / data?

Level of infrastructure build-up as we transition from low to high urbanization

- Benefits the same for different urban environments?
  - NO – drivers of different ages, cost prohibitive (expensive automated vehicles – not everyone will have one?)
  - Depends on the model on who owns the vehicles, who deploys it?
  - Transit vehicles that can provide mobility for low income
  - Early adopters benefit more? Where do they reside – urban & sub-urban?
  - Fully automated vehicles can end transit (light rail systems)?
    - Afford smaller vehicles, ride share and cut costs significantly
    - All transit systems are heavily subsidized at the moment
    - Can bring about changes in culture – healthier living by providing more options for people?

III. Day 2 Presentations and Discussion

Opening Remarks

The opening remarks described the two discussion topics for this session. The first topic focused on traffic management and operations with driverless vehicles (NHTSA level 4 automation). The second topic focused on the needs for modeling and simulation of traffic management and operations with automated vehicles.

Panel/Presenter Summaries

The group discussed how driverless vehicles could change public transportation, transit and shuttle operations, demand responsive transport (cyber cars), freight transport applications and infrastructure requirements.

Discussion Topic 3

Speakers	Summary of Presentation, Question and Answer Session
Edward Fok (Federal Highway Administration)	Mr. Fok presented a thought provoking discussion on arterial operations with driverless vehicles. His presentation touched various aspects of arterial management and focused on the external human machine interface which includes interaction of vehicles with pedestrians and vehicles with bicycles. He also mentioned failure modes such as low saturation condition and platoon separation dynamics.
Shannon McDonald (Southern Illinois University Carbondale)	Ms. McDonald introduced how different aspects of parking will be affected by various levels of automation. She mentioned that automated vehicles had the potential to change urban dynamics by possibly moving parking centers out of the city center allowing more room for economic activity.
Corey Clothier	Mr. Clothier presented how a phased approach will be essential to introduce

(Comet Robotics)	automated vehicles. He presented various projects going on around the United States that has deployed automation at various levels. These deployments are mainly low speed but have proved effective and efficient. He emphasized the importance of last mile automated transit pilots.

Summary of Discussion Topic # 3

Discussion Topic 3 "Management and Operations of Driverless Vehicles (Level 5)"

Freeway and Arterial Operation

- Automated cars – dynamic warnings (disabled cars, etc.).
  - Heavily instrumented
- Automated vehicles are going to know the geometry
  - Can also eliminate glares from the rising / setting sun
- Infrastructure spending?
  - In-vehicle apps might reduce spending on infrastructure
  - Fail safe / fail soft apps?
  - High speed travel might limit sensor range?
  - Might need support from additional infrastructure
  - Safety of infrastructure – hacking, vandalism, etc.?
  - Might have a higher speed limits
  - Based on support from infrastructure data to augment range of existing sensors

Public Transport, Transit, Shuttle Operation

- Can we separate trucks and cars on existing infrastructure
  - Studies are going on
  - Grade separation but can we do digital separation?
- Also need to think about pavement quality
- Major problems arise when trucks change lanes or merges.
- Difficult negotiation with labor unions as automated transit could affect existing jobs?
- Revised performance measures for transit operators
  - Number of passengers versus on time?
- Reduced wait times can have an effect on number of people preferring transit
- Equity issues –> Security challenges
- Transportation socio-equity problem for Atlanta

Freight Management

- Smart and dynamic routing
  - Increase value
  - Can change vehicle forms
  - And gives drivers a chance to do other things
- Similarity drawn to automated farm equipment
- Dramatic developments will take place first with fleets – business case
- Local distribution of goods can improve substantially with automation
  - Change in size of freight trucks

  - Break up weights to navigate through several states with different weight rules
  - Logistical changes
- Challenges
- Incomplete map databases

Discussion Topic 4

Speakers	Summary of Presentation, Question and Answer Session
Bart Van Arem (Technical University Delft Transport Institute)	Mr. Van-Arem first mentioned how current traffic flow theory uses typical human capabilities for modeling traffic (Car following, multi-anticipation, delayed response and hysteresis). Then he focused on how automation will change these aspects of traffic modeling since the human is taken out of the equation. He also emphasized how connected automation is more important than just automation. He left the audience with this question: What new paradigm does connected automation bring to traffic modeling and simulation?
Jamie Lian (Pacific Northwest National Laboratory)	Mr. Lian presented challenges and opportunities related to control of fully connected transportation system. Mr. Lian discussed the string stability issues in the platoons of advanced cruise control systems. In addition, he showed the difference of having traffic control based on V2V or V2I.
Ismail Zohdy (Booz Allen Hamilton)	With in-vehicle automation and vehicle connectivity gaining momentum, CACC systems are expected to enter the market in the near future. Dr. Ismail Zohdy presented the potential benefits of optimizing vehicle trajectories approaching an intersection to minimize the total passenger delay. The proposed framework attempts to manage the movement of autonomous vehicles to prevent crashes while minimizing the total passenger delay, simultaneously. This concept is similar to the Transit Signal Priority (TSP); however, the APP concept can accommodate more complex situations, different maneuvers, weather conditions, and levels of market penetration of vehicle automation.

Discussion Topic 4 "Modelling and Analysis of Automated Vehicles"
Tools to accommodate automated vehicles (micro and macroscopic models)

- Collaboration among different departments / agencies / countries is important for modelling
- Should pay attention to connected vs automated when modelling
  - Assumption automated vehicle will be connected (NHTSA mandate)?
- Challenge is the mixed environment and partial automation (in the next 10-20 years)
  - No base case available to use as a starting point
- Level of modeling difficulty depends on how connected communications are deployed
- How far off will we be if we use existing models when modelling automated vehicles?
- Pedestrians and cyclists are increasing in cities and this change might be independent of how automation progresses.
- Need to distinguish microscopic and macroscopic models
  - Different approaches to validate the models
- Key requirement for research is to have data
- New forms of driving behavior
  - If people are given the option to turn on and off systems – Need to understand when do they do that?
- Need a multi-disciplinary approach

- Is computing power a concern?
  - Should not be an issue with current technological advances
  - Including communications could double need of computing power

IV. Key Results

- Automated vehicles expected to increase traffic flow efficiency
  - Connected automation required
- Vehicle as a sensor and traffic control actuator
  - Regulations, laws?
- Connectivity can bring early benefits, automation can leverage them
- Traffic management context will change
  - Travel/logistics patterns, types of vehicles, parking, empty cars
- Transition period is important
  - Mixed traffic, manual-automate transitions, managed lanes
- New management strategies needed
  - Centralized, decentralized, hierarchical, transactional
- Driverless vehicles will require new management approaches
  - Intersection control, platooning
  - Management of empty vehicles, parking
- Dedicated lanes may accelerate deployment of driverless vehicles
  - Tracks in pavement, equity issues
- Lack of suitable modeling and simulation tools to accommodate automated vehicles (all levels)
  - Data, calibration, control models, information flows, driver behavior
- Traffic flow dynamics will fundamentally change
- Adapt from other disciplines (agriculture, power)

V. Next Steps

- The audiences for this topic will be asked to provide further feedback to the "prompt" questions posed in this breakout session
- The organizers will prepare a summary for follow up at the 2015 TRB Annual Meeting

Appendix C: Truck Automation Opportunities Proceedings

Session and Proceedings Organizers: Mohammad Poorsartep (Texas Transportation Institute), Alain Kornhauser (Princeton University), George Munteau (Pacific Northwest National Laboratory), Thomas Stephens (Argonne National Laboratory), John Estrada (Dering & Estrada), *Brian Routhier (FMCSA), Osman Altan (FHWA), Edward Fok (FHWA), Deborah Freund (FMCSA), Keith Kahl (Oak Ridge National Laboratory)*

I. Session Focus and Goals

The Truck Automation Opportunities breakout session focused on the discussion concerning how automated vehicle technology could affect commercial vehicle operation. It is important to note that commercial vehicles represent the transport of goods and services (e.g., truckload operations, less-than-truckload) and are covered by strict regulations. Myriad policies and regulations for commercial vehicle operation could be impacted by more advance levels of automation that are being introduced to the market. Hence, it becomes inevitable for the industry to consider and plan for possible upcoming changes while the technology moves along its maturation trajectory. This session revolves around the challenges and opportunities that will be presented from the perspectives of different stakeholders.

II. Day 1 Presentations and Discussion

Opening Remarks
Mohammad Poorsartep

- Provided a recap of TRB Automated Vehicle Summit, 2013 Truck Automation Session and highlighted the several main areas of discussion namely:
  - Business Case
  - Technology,
  - Policy and Human Factors

The session then proceeded with other speakers, further discussed below.

Panel/Presenter Summaries

- Day one of this session focused on the perspectives presented by government entities, namely the U.S. DOT, U.S. DOE, and DoD. However, the representative from US Army United States Army Tank Automotive Research Development and Engineering Center (TARDEC) was unable to provide the audience with the latest update. Hence, the focus was put on U.S. DOT and U.S. DOE activities.

Speakers	Summary of Presentation, Question and Answer Session
Mohammad Poorsartep, Project Manager, Texas A&M Transportation Institution *Overview of 2013 Truck Automation Session:*	• Presented an brief snapshot of what was discussed in the 2013 which focused mostly on three topic areas o Business Case o Technology o Policy and Human Factors • The presentation summarized the top 6 research topics identified: o Truck automation applications cost benefit analysis o Components of automated commercial vehicle training o Factors of dynamic optimal platoon ordering o Faults, failures, and failovers

<table>
<tr><td></td><td>
<ul>
<li>o Driver skill degradation with automated commercial vehicle operators</li>
<li>o Smart truck and dumb trailer challenges</li>
</ul>
<ul>
<li>He then proceeded with the 3 topics that were down selected by attendance in order of importance
<ul>
<li>o Truck automation applications cost benefit analysis</li>
<li>o Factors of dynamic optimal platoon ordering</li>
<li>o Components of automated commercial vehicle training</li>
</ul></li>
</ul>
</td></tr>
<tr><td>Jeff Gonder, Section Supervisor, National Renewable Energy Lab

NREL's Truck Platooning Test Results and NREL's Available Data</td><td>
<ul>
<li>Provided an overview of data sets and their characteristics available at National Renewable Energy Lab (NREL) that could support truck automation. Following is a list of available such data sets collected:
<ul>
<li>o Alternative Fuels Data Center (AFDC)</li>
<li>o National Fuel Cell Technology Evaluation Center (NFCTEC)</li>
<li>o Transportation Secure Data Center (TSDC)</li>
<li>o Fleet DNA Data Collection</li>
<li>o FleetDASH</li>
</ul></li>
<li>He then further described the Fleet DNA data sets that can be an asset for those working on automation. Fleet DNA Captures and quantifies drive cycle and technology variation for the multitude of medium and heavy duty vocations</li>
<li>This presentation then continued with a discussion regarding the recent Class 8 Tractor Trailer Platooning testing that was done by NREL.
<ul>
<li>o Team fuel savings ranged from 3.7% to 6.4%</li>
<li>o Closer following distances caused the engine fan on the trailing truck to engage, negatively impacting fuel savings</li>
<li>o Lead truck: 2.2% to 5.3% savings @ 65,000 pound gross vehicle weight</li>
<li>o Shorter following distances consistently produced greater fuel savings</li>
<li>o Trailing Truck: 2.8% to 9.7% fuel savings</li>
<li>o Tests with no "fan on" time had savings of 8.4% to 9.7% for trailing truck</li>
<li>o Significant line-haul fuel savings possible through platooning</li>
<li>o Engine coolant temperature needs to be monitored/addressed for the trailing vehicle</li>
</ul></li>
</ul>
</td></tr>
<tr><td>Osman Altan, Senior Transportation Specialist, FHWA, U.S. DOT

Overview of FHWAs Truck Platooning Project</td><td>
<ul>
<li>Provided an overview of two related projects that are currently underway and funded by the U.S. DOT:
<ul>
<li>o 1- Partial Automation for Truck Platooning – "Assess the Feasibility of Deploying Partial Automation for Truck Platooning"</li>
<li>o 2- "Heavy Truck Cooperative Adaptive Cruise Control System, Testing and Stakeholder Engagement for Near-Term Deployment"</li>
</ul></li>
<li>Both projects have platoon implementation on trucks followed by demonstration</li>
<li>The first project focuses on answering if three-truck CACC performance is achievable in mixed traffic, driver preferences for CACC time gaps, energy savings at preferred time gaps, benefits in truck lane capacity, energy and emissions, and deployment strategies for truck CACC</li>
</ul>
</td></tr>
</table>

	• The second project is aimed at finding answers to two-truck platoon operation in mixed traffic, platoon operation with non-identical vehicles, human factors, fleet operations acceptance, and system robustness • Osman also presented the results from a study conducted by the American Trucking Research Institute (ATRI) under the second project that are as follow: o Both groups are comfortable with technology o 54% of carriers and 40% of drivers unsure about safety impact o 38% of carriers and 18% of drivers expect reduction in crashes by this technology o Will be deployed with larger regional/long haul carriers with more trucks in their fleets o Couple with their trucks rather than competitors' vehicles o 54% of drivers unwilling to use the system (carriers think 44% unwilling) due to discomfort o 44% of drivers maintain a following distance of 6 seconds or more
Deborah Freund, Senior Transportation Specialist, Federal Motor Carrier Safety Administration-U.S. DOT *Impacts of automation on safety policies and regulations*	• Provided a thorough overview of the Federal Motor Carrier Safety Administration (FMCSA), its practices, jurisdiction and more. • She emphasized soft failure, robustness etc. as many of these systems are secure and packaged, but have to be tested and maintained. • She expressed that regulations related to drivers depends on level of automation. Levels 0 through 3 may not need any driver related regulations as driver is expected to be ready to control but level 4 is not quite clear yet. • Safety is at the forefront and they may need to consider different inspection models. • Compliance, Safety, Accountability (CSA) does not provide credits for the use of safety systems. Other entities might view it differently. Motor carrier insurance companies are asking for discounts. Do not have statistics for claims picture.

Summary of Other Facilitated Group Discussions

[DISCLAIMER: Summaries presented here are not necessarily the views of the presenter. This section is compiled based on the discussions that took place during speaker's presentation and Q&A session]

Discussion following Mohammad Poorsartep presentation:

- Interoperability will be important to allow different trucks to form, join and leave platoons, since the same trucks will not make up a platoon across the country. How can trucks find each other (between different carriers), and form integrated platoons?
- Potential implications of transfer of control, workload transition. Can't have a sleeping driver wake up and be expected to take control immediately.
- FedEx, for example, doesn't own the tractors used, and can't require what kind of truck to be used. Might be able to incentivize adoption of technologies of there is a benefit.

- Who realizes savings from using less fuel, driver or carrier? What is the model for distributing the benefits fairly among platoon participants?
- Many small fleets, not only independent owner/operators, but also drivers operating under lease agreements.
- Only about 25% of class 8 tractors are sold with stability systems, even though these have been available. Some suppliers offer as standard, but often optional. Truck purchasers often choose not to buy optional technologies, even when the options are not very expensive. Government will have a critical role in pushing adoption (or requiring by regulation).
- If we have different vehicles with different integrators, string stability becomes quite complicated with standardization. Have to be within tight tolerances. For example, how do you measure acceleration?
- There are 5,000 motor carriers that FMCSA knows about. Over 90% operate with 6 or fewer power units. 80% of these operate under lease agreements to larger companies like FedEx.
- Adoption and acceptance subtle and important: experience with self-driving race car. Trepidation with people approach. After the first lap, people relax. Initial hesitation is much larger; acceptance is much larger than it should be. Bimodal characteristic of fear followed by shallow acceptance.
- The human side is another consideration. Mental health issues and anxiety depression, etc. have a high incidence among truck drivers, which is not surprising since truck driving so isolated. Platoon may increase social aspects. What human issues can come out of this? Big benefits that can't be quantified.

Discussion following Jeff Gonder presentation:

- Data collections activities and databases that are available through NREL and could be used by the industry in developing new technology: AFDC, NFCTEC, TSDC, Fleet DNA (med- and heavy-duty drive cycle and powertrain data), Fleet DASH (federal fleet).
- The Fleet DNA program may be accessed here: www.nrel.gov/fleetdna
- Field evaluation projects: collect data on use of hybrids (hydraulic hybrids), natural gas, fuel cell bus, Berks Area Regional Transportation Authority inductive charging, etc.
- Data processing: standardizing format, quality control, link to reference data (road network, grade, GIS layers), meteorological, economic, land use, vehicle and demographic information
- Analysis tool: DRIVETM (Drive-cycle Rapid Investigation, Visualization and Evaluation Tool)
- Drive cycle data. Statistical information on drive-cycle data is publically available. Second-by-second data aren't publically available.
- Truck platooning testing with Peloton was conducted on 0.5 mile oval test track in Uvalde, Texas in accordance with SAE J1321 Type II fuel consumption test. Trucks were American-style line haul sleeper cabs with modern aerodynamics (SmartWay, trailers with side skirts). Tested up to 70 mph at various loads.
- DSRC, radar and laser vision, vehicle braking and torque control interface were installed on each truck. Ten constant speed tests and one variable speed test. 20-75 foot vehicle gaps with two to three reps per test, 60 miles each rep and data were collected using on-board data collection systems. Trucks tested after warm-up. Baseline (non-platooned truck) run. Platoons of two trucks run simultaneously with control (isolated) truck.
- At 65 mph, 65,000lb, up to 9% fuel savings (up to 5.3% lead truck, up to 9.7% trailing truck, 6.4% for two trucks), even at gaps of 75 ft. Counterintuitively, trailing truck showed less fuel savings at low following distance, since belt-driven fan had to run. With heavy payloads, fuel savings were slightly lower.

- In the future, truck designs may mitigate air cooling/heating fan concerns.
- With more trucks in a platoon, middle trucks realize even more fuel savings. Smaller marginal benefit and more operational complexity with more trucks in a platoon.
- ATRI and Auburn studying fleet data to estimate opportunities for platooning.
- How well aligned (laterally) does following truck need to be before benefit is decreased. Curves can affect aerodynamics and fuel savings.
- In California PATH testing of three-truck platoons, second truck was offset 6 inches (to avoid interfering with communications between first and third truck).

Discussion following Osman Altan presentation:

- Osman provided an overview of two current truck automation projects funded by U.S. DOT: "Assess the Feasibility of Deploying Partial Automation for Truck Platooning" and "Heavy Truck Cooperative Adaptive Cruise Control System, Testing and Stakeholder Engagement for Near-Term Deployment".
- The first project seeks to answer the following questions: Is platooning performance achievable with truck CACC in mixed traffic? Driver preferences for CACC time gaps? Energy savings at preferred time gaps? Benefits in truck lane capacity, energy and emissions? What are the deployment strategies for truck CACC? What are the synergies with I-710 truck lane development?
- The first project will also investigate operational issues such as: Planning of platoon formation using DSRC, Range extension by 3G or LTE wireless technology, Truck sequencing, Platoon formation, financial transactions among trucks, Loading/weight of individual trucks/trailers, Human factors, Driver-Vehicle Interface and Information to drivers.
- The second project will investigate topical areas in: platoon operation in mixed traffic, platoon operation with non-identical vehicles, human factors, fleet operations acceptance, system robustness (i.e. communication disruptions and sensor errors).
- U.S. DOT has funded another study to collect naturalistic truck following behavior. The data collected are expected to assist with development of automated truck platooning applications.
- The latter study will provide data to address the followings: How closely do trucks follow other vehicles on highways? How does following behavior vary by different factors (e.g. type of truck, speed, road type, road conditions, weather, visibility, time of day, etc.)? At what distances do vehicle cut-ins occur? What is the safety impact of different following gaps? Which elements of truck following behavior should be considered in the development of automation technologies, specifically platooning applications?
- It was emphasized that adding advanced technologies should not have a negative impact on truck down times.
- Platoons may cause structural degradation for bridges due to tight following distances or have a negative impact on pavements as two or three heavy loaded trucks follow almost the exact footprint of the leading trucks and wearing out a narrow path.

Discussion following Deborah Freund presentation:

- Deborah provided a thorough overview of FMCSA, its operation, regulatory authority and some statistic that showed the depth and breadth of this department. FMCSA is relatively small and was stood up in Jan 2000 as part of Federal Highway Office of Motor Carriers. However, changes in legislation made it necessary to have this organization on the same level as other departments within DOT.
- Even if a business is providing another service, they are considered motor carrier. FedEx UPS, Swift, US express, but also US foods, petfoods, PepsiCo, private motor carriers that ship their own goods.

More private motor carrier than commercial for hire 10.6 million large trucks and over 764 large buses are currently in use in US. 3.5 million roadside inspections were conducted in 2013.

- FMCSA can regulate drivers in terms of licensing of larger vehicles and 16 or more passenger vehicles and hazardous materials. general physical, drugs, hours of service
- FMCSA also has regulatory authority over vehicles and that includes parts and accessories necessary for safe operation, In US no minimum vehicle size is required to be considered as motor carrier.
- Question was raised whether FMCSA will regulate the following distance of platooning trucks.
- Deborah also mentioned of two Small Business Innovation Research projects that are funded by FMCSA: 1) Automated Trailer Vehicle Identification Number Identification and Sequencing System. 2) Determination of Towed Unit (Trailer) Characteristics from within the Powered Unit (Tractor).

III. Day 2 Presentations and Discussion

Opening Remarks

- Mr. Poorsartep opened the session by a quick recap of Day 1 presentations and discussions which focused mostly on activities currently underway and sponsored by the U.S. DOT and the U.S. DOE.

Panel/Presenter Summaries

- The second day of the Truck Automation Opportunities session, unlike the first day, focused only on the perspectives presented by the private sector. These included representatives from the fleet industry, truck manufacturers, and suppliers, speaking to their current and future needs, challenges, and opportunities in the future of automated trucks.

Speakers	Summary of Presentation, Question and Answer Session
RJ Cervantes, Director, California Trucking Association *How Fleets Adopt Advanced Technologies*	• RJ began focused his presentation on the characteristics of the trucking industry, how it operates, what operators expect from advanced technologies, and how policies impact (have impacted) adoption of new tech in the trucking industry. • He provided some interesting stats such as in CA, more than 80% of tonnage handled by trucks and more than 80% of communities depend exclusively on trucks. Trucking industry views automation as 15-20 years down the road. • 50% of fleet have 3 or fewer trucks. Nationally, according to U.S. DOT, 90% of fleets operate 6 or fewer trucks, 97% fewer than 20. • Trucking industry in general is currently more engaged and interested to hear about levels 1 and 2. • 50% of all trucks driven are owned by owner-operators working mostly in ports and agriculture sector. A major focus is currently hours of service regulation by FMCSA. Drivers can do 70hrs/week and truckers don't make money when they're not moving which causes concerns regarding level of productivity • Labor and fuel cost are the most important concerns. Diesel price is much higher in California. 12-15cents per gallon increase is possible while state guarantees a minimum 5 cent increase per year.

Fred Andersky, Director, Bendix *Suppliers View of Truck Automation and the Road Ahead*	• The presentation captured the perspective of the supplier community, their challenges and opportunities that they see lay ahead. He also provided some stats on truck related accidents that can and has been driving the implementation of several safety related technologies. Though accidents have decreased for the last ten years, there has been an increasing trend over the past few years. In 2012, there was a 10% increase in truck related accidents. Fatalities were up by 4% and injuries up by 18%. • He pointed out that beyond safety reasons, automated trucks can provide environmental and operational efficiency benefits. These improvements could range from expected fuel savings of 5-20%, Driver related cost savings of 35-100%, or fleet productivity increases of 25-30%. • In terms of safety benefits, insurance costs related to level 5 autonomous vehicles are expected to lower by 20-60% or safety performance for fleets, such as CSA scores, improvements are some examples related to safety aspects of the technology. • Due to their weight, commercial vehicles (class 8) can have 13x-27x more kinetic energy of a passenger car that imposes certain requirements when designing for automation. • He introduced the safety systems 4I's concept (i.e. Information, Intelligence, Intervention, Insight). Each of these elements will need further improvements to be ready for autonomous vehicle technology. For instance, enhanced cameras & radars; LiDAR; integration; connectivity, object and situation arbitration; predictive, much stronger declaration & steering intervention, telematics, data for driver education; maintenance; traffic control; analytics. • He also discussed limits to today's conventional advanced driver assist systems today that: can't predict driver skill, can't predict level of friction on road surfaces ahead, can't predict traffic, can't look around corners , Can't look beyond the range & dynamics of sensors, can overwhelm/distract driver. Fred believes these limitiations cannot be overcome with mid-term available individual sensor techology. Senors & system information of other traffic participants and infrastructure are required. • He pointed out that the evolutionary path for truck automation start with platooning followed by autonomous control of select maneuvers / functions such as (Controlled, monitored lane change, executed by the system per driver demand , automatic rear docking & coupling , automatic vehicle positioning in a lane, auto-pilot on select road stretches). • Fred noted that driver assistance systems today and in the future do not replace the need for a safe driver practicing safe driving habits and comprehensive driver training. Drivers remain responsible for safe operation of the vehicle at all times.

Dan Williams, Chief Engineer, TRW *Steering Solution in Support of Truck Automation*	• Provided a thorough overview of electric steering systems that could be used for truck automation applications. His presentation also covered the differences between commercial and passenger vehicle markets. • From his perspective, passenger vehicles appeal to safety and convenience. But we know drivers get significantly better through experience. Insurance statistics tell us drivers improve greatly in the first two years of driving. Hence, "level 3" is likely "an accident waiting to happen" as autonomous controller cedes control back to an untrained, disengaged and increasingly inexperienced driver in unusual circumstances. • On the other hand, commercial vehicles, appeal to vehicle utilization through fuel economy and throughput efficiency. • Two distinct portions of commercial vehicle duty cycles: 1) High speed lanekeeping, 2) Low speed vocational maneuvering. Both of these modes are ideal for automation in ways that general passenger car applications are not. • For commercial vehicles, lane maintenance mode uses the most fuel and is most easily automated and this could possibly be an area to benefit from automation. Also, vocational maneuvering requires skilled operators and can result in equipment damage. This application will definitely require lateral control. • Transit bus duty cycles are completely different from over-the-road motorhomes, dominated by low speed/high wheel cut maneuvers. Work ratio is the ratio of driver work to steering gear input work. This metric has been shown to have convenient invariance to route. Torque overlay reduces transit driver workload by70%. Low efforts create distinctive value hence creating value for electrical steering systems. • Artificial torque can effectively transmit vehicle dynamic information to driver.
Joakim Svensson, Director, Volvo Trucks *Volvo's Perspective on Truck Automation Development*	• Volvo's perspective on truck automation presented by Joakim focused on the views of OEMs on how truck automation could move forward and the challenges and opportunities to see in implementing truck automation. • Volvo Group related business areas are Volvo Trucks, Volvo Buses, and Volvo Construction Equipment. These three different sectors imply two ways with lots of synergies i.e. controlled environments and public roads. • What drives commercial vehicle automation is the fleet operating costs. The major cost categories are fuel, driver, and truck/trailer. These three categories comprise of nearly 75% of total operating cost. • Volvo currently has the following active safety systems in production: ESP, FCW-AEBS, ACC, LCA, LKS, and VDS (Volvo Dynamic Steering). • Joakim provided a comparison of truck platooning results achieved from Jari and SARTRE projects. • From his perspective, CACC platoons are starting to become mature for customer pilots. • Platooning (CACC) can be deployed within a truck fleet, requiring most likely a multi-brand solution. Another deployment path is to have ad-hoc

		platooning, i.e. platoons are formed on the fly on the road. Then some specific factors need to be considered regarding truck platooning: 1) platoons are dynamically formed, dissolved, and reformed though a platooning system" 2)Trucks do not have to be from the same logistic company 3)There needs to be a "platoon system" owner.

Summary of Other Facilitated Group Discussions

[DISCLAIMER: Summaries presented here are not necessarily the views of the presenter. This section is compiled based on the discussions that took place during speaker's presentation and Q&A session]

RJ Cervantes

- Due to legal barriers related to employment and litigations, the intermodal operator model is thriving. Truck driver industry has shifted from long haul towards smaller trucks, using rail for long distance. This is due to labor costs, driver shortages, fuel cost, and myriad other concerns. This way they can get people home at night, under service hours, and bring their operation cost lower.
- Trucking industry is a cutthroat operation. After deregulation in 80s and 90s anyone with a phone and $10,000 can start a trucking business, which caused a large spike in number of carriers. The FMCSA CSA scoring aims to reduce these less safe carriers which puts pressure on consumers as it increases the cost.
- Misclassifying workers is a major concern (i.e. independent contractor vs employee) for carriers and it materialized when CA Resource Board mandated new equipment (20,000 in emission reduction or 130,000 trucks etc.) and this forced many out of business. Drawing from smaller labor pool in ports did hurt industry in long run. States have heavily focused on misclassification claims.
- Regarding automation, policies and regulations have to be flexible: owner operators move agriculture products, dirt and aggregate, seasonal commodities, hauling containers, to rural places, etc. "Can you take it to fields in central California for tomatoes?" Owner-operators need maximum flexibility for switching.
- The trucking industry is one of the few remaining blue-collar industries open to those with only a high-school diploma. What are the political risks? It may be difficult to convince labor pool. Trucking is still an option for entrepreneurial types. Industry and automation groups need lobbyists.
- Question was raised regarding dedicated trucks lanes and RJ's response was CA facing infrastructure problem. Two thirds of the infrastructure is in poor/mediocre condition. Proposition 1B bonds has expired which means loss of 3B infrastructure revenue annually. So the fear is how to pay for such dedicated truck lanes? Until then, truck lanes will not become a reality.
- The question was raised, "Given that most trucks are owner-operator and on the other hand, there is a limit on how much each driver can drive per week. What is happening to these trucks? Are they sitting idle?" Yes. That's the problem. Even in larger fleets, it is a problem due to labor shortage.
- Industry's shift from long haul to short range is a concern. For platooning system it will pay off for long distances and high speeds. But if the logistics industry is moving away from long haul operation, platooning may no longer be feasible. This comment was responded by the fact that if fuel and labor cost continue to increase, but this technology could offset the increase, carriers may come back and re-consider long haul operation, instead of using rail.

Fred Andersky

- The truck is a factor of production for the fleet/owner-operator. Therefore, cost of the vehicle is critical in the decision to purchase. Added equipment needs to justify its value – deliver an acceptable ROI and payback period (typically within 2 years.). ROI on safety is important. Many fleets are self-insured, so accidents can impact their bottom line. With profit margins low around 3% a crash costing $20 million can be a disaster depending on fleet revenues.
- When antilock braking systems were first available, only around 30% of the market was buying it despite its benefits. Legislation proved beneficial by in helping with full adoption of the technology
- The question was raised whether owner operators or fleets are more open to these technologies. Fleets tend to be bigger proponents of tech. They have capital available for tech while owner operators often buy second hand trucks.
- On the fleet side, anything that prevents being pulled over is worth considering such bypass way station, self-inspection, etc. to improve efficiency as idling hurts fuel economy. However there are privacy concerns and fears.
- FMCSA is conducting field tests to show the impact of heavier trucks on braking systems—today out-of-service at 20% of brakes not functioning. Also, reduced stopping distance (RSD) regulation performance can be impacted when brakes are relined with friction materials not designed to RSD specifications. To accomplish RSD performance, brake manufacturers developed new systems involving larger drums, different frictions and shoe materials. While NHTSA regulations impact new vehicles, there are no aftermarket requirements for ensuring friction meets the regulation. Some popular replacement brake linings may not work as well, impacting stopping distance performance.

Dan Williams

- Not quite sure if lateral control is required for platooning to realize fuel saving benefits.
- Most drivers are conditioned to push their vehicles slightly to the right, due to incoming traffic from the opposite late, even though incoming traffic may not exist or separated by median.
- One possible reason why electric steering assist systems have not moved into the commercial vehicle space and are currently only implemented in motorhomes is due to high prices.
- In Europe it is easier to get insurance benefits related to advanced safety systems. More difficult in the United States, because actuarial tables are needed first to prove out the technology and insurance related benefits.
- Automated manual transmission faced strong resistance by the industry twenty years ago. However, new drivers do not like to shift. Behaviors change over time, and same will be applicable to automated steering and other similar technologies.

Joakim Svensson

- Reverse driving related accidents are very common among heavy duty vehicles.
- Drivers are never trained in "near-accident" situations! This could cause problems for any average driver if they are faced with a near accident situation since they don't know how to react. This is in contrast to the aviation industry where pilots are trained on how to react and deal with near accident situations.
- There needs to be pilot deployment of truck platooning to measure the actual benefits of truck platooning as opposed to potential benefits (current literature only measure possible/potential savings and benefits).
- It took about 10-15 years to go from cruise control to ACC to Lane Keeping Systems plus ACC. How long will it take to get to CACC and full control platooning?

- There needs to be a comparison between European and American platooning tests that were conducted. Results are not quite similar and there is some insensitivity to following gaps that were seen in some tests, but not in other tests. (For instance, use two trucks formation, similar environmental condition, etc. and compare European trucks with American trucks – rigid trucks, vs long nose vs etc.)
- There was a discussion regarding whether we should wait for V2V market penetration to catch up in order to enable platooning, or is it possible to deploy platooning application without V2V for two trucks driving in tandem. No agreement was reached in this discussion which raises the question if there are other alternative to DSRC based V2V to enable two trucks platooning.
- There needs to be studies focusing on what following gaps are comfortable for drivers. The gap that may provide the best fuel efficiency may cause discomfort for the driver. There needs to be study to find the sweet spot.
- Also, there needs to be a study to determine, what the minimum required gap is, given the latency of sensors and V2V to ensure the safety of the platoon.

IV. Key Results

- Discussion took place regarding whether we should wait for V2V market penetration to catch up in order to enable platooning, or is it possible to deploy platooning application without V2V for two trucks driving in tandem. No agreement was reached in this discussion which raises the question if there are other alternative to DSRC based V2V to enable two trucks platooning.
- Key questions regarding implementation of CACC truck platoons that need to be answered are:
    - Driver - system HMI a key factor for user acceptance
    - Operating conditions
    - Dependability & liability
    - Mixed Fleets
    - How to form the platoon?
    - Can it be stepwise deployed?
- A need exists for studies focusing on following gaps. These studies include:
    - What following gaps are comfortable for drivers? The gap that may provide the best fuel efficiency may cause discomfort for the driver.
    - What the minimum required gap is to ensure the safety of the platoon, given the latency of sensors and V2V.
- A number of open questions remain:
    - What do truck fleets specifically need or want in terms of automation (e.g., dedicated infrastructure for level 3 automation and higher)?
    - How are owner/operator needs and challenges different from fleets in regards to these technologies?
    - What steps are necessary to alleviate other barriers to technology introduction such as legal, societal, or user perception?
    - What is the business case for each level of automation?
    - What are the upfront, ongoing, or hidden costs of commercial vehicle automation?
    - What is the incentive/disincentive? For example, are there insurance savings, decreased labor costs, or regulatory benefits?
    - What are the risks of early adoption?
    - What is the ROI where the owner of the truck is also the driver?
    - What infrastructure is needed for truck automation?
- Political considerations exist (e.g., Will automation eliminate good paying blue-collar jobs?)

V. Next Steps

- Demonstrate that the technology works and is reliable
- Conduct pilots with customers, to verify potential market for CACC and potential fuel savings
- Create standards
- Identify and eliminate barriers (authorities, legal, market)

Appendix D: Legal Accelerators and Brakes Proceedings

Session and Proceedings Organizers: Erin Flanigan (Cambridge Systematics), Karlyn Stanley (RAND Corporation), Ellen Partridge (U.S. DOT Office of the Secretary), Frank Douma (University of Minnesota), James M. Anderson (RAND Corporation), Tinu Diver (U.S. DOT Volpe Center), Francine Steelman (Miami-Dade Expressway Authority)

I. Session Focus and Goals

This breakout session explored the legal issues that might accelerate or inhibit the deployment of autonomous and connected vehicles. The first two panels focused on different transportation modes, and the third discussed the regulatory environment in which these modes must function. Throughout the two days, participants were asked to focus on core questions relating to: (1) what mode is most likely to lead deployment in the current regulatory environment; (2) what role will "infotainment" play in driving or delaying deployment; (3) whether uniform laws would help or hinder deployment; (4) what opportunities and threats exist from increasing data use, and misuse; and (5) how can liability questions be best handled. The final part of the session was a group discussion that sought to prioritize the most important questions and provide recommendations for moving forward.

II. Day 1 Presentations and Discussion

Panel 1- Infrastructure: Tollways, Highways and Transit

This session explored how public transit, highway, and toll authorities might use AV and connected vehicle technologies, and related privacy and data management issues. Will AVs mean changes in lane design and promote managed lanes? Will the data collected by AV's be available to these organizations? If so, how might they use it? Do AVs present new revenue opportunities? Will AV users be able to opt in or out?

Speakers	Summary of Presentation, Question and Answer Session
Moderator: Tom Bamonte, North Texas Toll Authority	
Francine Steelman, Esq. Miami-Dade Expressway Authority	Privacy and Data Management • Current data collection process via open road tolling • Live feeds, personal information not stored. • Only billing data are created and provided via customer accounts. • Miami-Dade Expressway Authority is strategically planning how to use driver location data to deliver a better service to their customers; to generate non-toll revenue. • We need to understand: What information is needed for a fully functioning in-vehicle system. Recent Supreme Court cases protecting data privacy: • *Jones*: Government used GPS tracking for 28 days. Court ruled that the 4th Amendment protects drivers from location tracking information. Therefore the government's access to that information required a warrant. • *Riley:* need warrant before searching contents of mobile phones • **Issue to watch**: balance of protecting public health, safety and welfare

<table>
<tr><td></td><td>against protecting personal data
<ul><li>Watch for public preference through "opt-in" choices</li>
<li>Public will agree to disclose personal data for a clear benefit; and if information use is limited and not punitive.</li></ul>
Questions:
<ul><li>The 4th Amend restricts governmental use of private information; what about private industry?
<ul><li>Private industry will lead the charge. As time goes on and people buy connected vehicles, they will "opt-in" to sharing their personal data. Interpretation of the 4th Amendment will expand in accordance with the public's changed expectation of privacy regarding personal data.</li></ul></li></ul></td></tr>
<tr><td>Coco Briseno, Caltrans</td><td><ul><li>1997 Automated Highway System was a "wow" lots of excitement and energy, but look where we are today (regrouping, shifting to AV).</li>
<li>Considerations of infrastructure construction for DOTs.</li>
<li>Legal, societal, and institutional issues were the greatest challenge, and drove the move from Automated Highway System to Connected Vehicles to Autonomous Vehicles.</li>
<li>Research benefits when we dust off work of past decades</li></ul>
Questions:
<ul><li>What about the truck platooning corridor?
<ul><li>Currently undergoing a statewide freight plan and there is talk about dedicated truck lanes, status is unknown, but the issue is being looked at.</li></ul></li>
<li>From state DOT Perspective what is your take on liability?
<ul><li>California is in the middle of the process of figuring out recently passed legislation going through rule making....Liability issues are right in the middle of these questions!</li></ul></li></ul></td></tr>
<tr><td>Jeffrey Spencer (for Vince Valdes), Federal Transit Administration</td><td><ul><li>Federal Perspective. What does the future look like?</li>
<li>FTA looking at mobility on demand and what the data needs are underlying this.</li>
<li>Eugene, Oregon VAA (vehicle assist and automation) had to stop testing because of liability issues (what about the bad actors? Hackers?) What are the penalties behind this?</li></ul>
Questions:
<ul><li>With budget problems in Washington, DC, how much of a break will this be on the AV program? Where will the money come from?
<ul><li>Lots of models to look at, such as private investment, the key is how to make private investment ubiquitous</li>
<li>What about equity and accessibility issues?</li></ul></li>
<li>How will we integrate these new starts into traditional planning approval process?
<ul><li>This is a challenge FTA is facing now and trying to work through. The Florida flexbus is an example of this.</li></ul></li></ul></td></tr>
</table>

Yeganeh Mashayekh, University of Pennsylvania, Carnegie Mellon	• Presented on the Pennsylvania DOT (PennDOT) 2040 vision for connected and autonomous vehicles • A 2014 – 2016 timeline was the focus. Looked at what PennDOT needs to do now? Findings were to start working on these things: o thorough evaluation of all existing and planned capacity/level of service (LOS) enhancements and ITS related investments; o collaborate with private sector to convert data into information; and o Prioritize safety and mobility applications. • **Challenge**: Only a third of all miles of roadway in PA are owned and operated by PennDOT.

Summary of Other Facilitated Group Discussion

Infotainment

Accelerator: More information (in any form) provided to the driver will increase drivers' distraction and thereby increase the need to automate the vehicle to increase roadway safety.

Brake: If increased infotainment is not coupled with vehicle automation then these distractions will create hazardous driving conditions.

Lead Role

Accelerator: Private business will be the leaders by integrating automated platooning technology into the transportation delivery systems. Use of this technology can increase the companies' profits by lowering the operating costs, increasing delivery efficiencies and better control/tracking of the transported goods.

Data Use and Misuse

Accelerator: Data held and controlled by private businesses should engender greater consumer trust that the data will be used for the limited purpose of providing services in the automated vehicle and this it will not be misused by the car maker. Otherwise, the consumer will choose another AV service provider that provides the necessary assurances; or the consumer will decide to not purchase an automated vehicle until those assurances can be provided.

Brake: Data that are held and controlled by a governmental entity without clear limitations and restrictions on the use and sharing of the information will create consumer mistrust in utilizing the automated vehicle technology. Question: Will law enforcement have access to this information?

Uniform National Laws

Accelerator: National and international laws are needed regarding automated vehicle technology to create consistency in the use of these vehicles; and to "normalize" the expectations of the world-wide consumers.

Liability

Possible Brakes: Will insurance companies have access to the information on the automated vehicle's "black box?" Can this information be used in a court of law against the driver?

Panel 2 – Vehicles: Commercial Trucking and Passenger

This discussion panel (no presentations) explored the ROI for the commercial trucking industry to adopt autonomous or connected vehicle technologies, whether managed lanes for platooning trucks would be necessary, and what role "infotainment" may have in driving deployment in both freight and passenger vehicles. Panelists were also asked to consider data questions, including how increased amounts of data may improve safety and efficiency and how these data can be stored and protected.

Moderator: Karlyn Stanley, RAND Corporation

Panelists

- Barbara Wendling, Principal Engineer, Volkswagen
- Alan Korn, Director of Advanced Brake Systems, Meritor Wabco
- Ali Maliki, Hybrid and Electronics, Ricardo
- Joakim Svensson, Director, Transport Analysis, Volvo

Summary of Panel Discussion

- The panel began by pointing out that true automation replaces the driving task – levels 3&4 (&5). While platooning creates this option for following drivers, the lead driver is still in control. Conversations about platooning and automation need to recognize this distinction.
- The trucking industry will employ technology that has payback. The question is what the payback is from level 3 or 4? Level 3 still has a driver, so no financial payback there – at the same time, safety pays. ROI drives the innovation. Fuel economy, productivity or improved safety all may contribute to improving the ROI, but we need figures to show financial benefit.
    - Platooning can save an incredible amount of fuel (5% for lead vehicle, 10% for following). These numbers would drive adoption. It is much easier to sell readily quantifiable numbers, like fuel economy. Safety is harder to quantify.
        - SARTRE showed use cases can be modeled for payback, and payment to lead vehicle (to cover lower returns). 4–16 m following distance needed.
    - The driver shortage is an issue: platooning allows lower cost, lower experienced drivers to follow (higher requirements for lead vehicle). 8–10% savings are estimated.
    - Legal issues:
        - Platooning fits the definition of tailgating. Need to change CA and other laws that make such short following distances illegal. (This is not just an issue for the US: European countries also have different minimum following distances)
        - How to prevent non-platoon tech vehicles from joining platoons? Need to certify platoon-tech equipped vehicles. Both to join, and to enforce against free riders.
    - Platooning is only a first step and would not be full time. Need to get practical results over full time automation.
- Since freight is continually seeking financial gain, they will be the first adopters–we need to find fleets that see benefit from day one, and get them to deploy.
- Different picture for passenger vehicles. No professional drivers, no quantified ROI. So more incremental roll out – adaptive cruise control, parking, braking long before platooning.

  - Current level 2 and 3 scenarios require a human driver to work. Closed campus and other shuttle type operations could be the first opportunity to eliminate human driver.
  - Infotainment will be a significant selling point to passenger customers. Traffic jam pilot frees up drivers to do other things, be happier, even more productive.
  - Infotainment as it applies to passenger, does not apply to commercial vehicles. However, telematics is huge for asset management and lowering cost per mile. It creates a new data source for fastest, shortest, most fuel efficient routes.
  - Telematics is integrated into operational systems now. Weather and traffic data will create better decisions.
- Data / Liability issues:
  - Recording driver attention and behavior will help determine liability. But, in Europe, filming outside the vehicle is surveillance, and illegal without consent of those being recorded.
  - Data are useful for accident reconstruction as well–especially in handling litigation. A new industry has sprung up for internal / external cameras in trucks.
  - Drivers must consent to in-vehicle cameras. It can be controversial, but good drivers know it can protect them.
  - Privacy issues are much more sensitive in passenger vehicles. Electronic data recorders (EDRs) are currently restricted to crashes and crash-like events. But on-going driver monitoring would have to be opt-in. It would never sell if it would be an option that makes it easier for them to get caught.
- Benefit from Uniform Laws?
  - Carriers in the United States are mostly national, and often global. Need to have minimum safety standards throughout the world, allowing validation and certification that works worldwide. Core minimum standards needed, but some local variation (licensing, etc.) would be acceptable.
  - Certain basic laws needed. Industry interoperability and safety standards in trucking may be enough.
  - Would be useful to have federal enabling legislation that will allow driving these vehicles across state lines without separate certification.
- Liability questions–who bears responsibility?
  - Liability concerns can kill a lot of projects. Also other legal issues (e.g., copyright in deploying hard drives). Industry will be paranoid about liability, but those issues will work themselves out.
  - Cybersecurity is greater concern. Internet is breached all the time, and it may not be possible to secure V2V. Therefore autonomous vehicles may be better option.
  - V2V is just warnings now–security becomes more important as V2V moves to directing vehicle movement
  - Concern about after-market deployment. Automaker most often the deep-pockets even if others made faulty equipment. Good lawyers protect automakers that way.

Summary of Other Facilitated Group Discussions

- Question: how to encourage innovation?
  - Develop, deploy and test to let best rise to the top. Other limits to deploying bad products (reputation, liability, publicity etc.).
  - Most regulations are minimum safety standards. Industry usually looks to exceed these as competitive advantage.
- How to prove tech is better than human?
  - Question will be how much better tech is supposed to be than human. Losing proposition is human as "fail-safe."
  - Anti-lock braking systems (ABSs) are an example of how to overcome this question. Fail-safe built in (when ABS fails, brakes apply).

- Response: even up to level 3, driver is supposed to still be alert. But if distracted driving laws exempt distracted drivers, then liability may shift.
- Depends on technology: drivers in following vehicles in platoons will not be able to respond if platoon technology fails. (Similar w/ level 4, i.e. if no steering wheel or brake pedal is provided).
- Level 3 and above will be so redundant and fail-safe they will not fail.

III. Day 2 Presentations and Discussion

Panel 3 – Government: State, Federal and International Regulatory Issues

Moderator: James M. Anderson, RAND Corporation	
Jesse Chang, Attorney-Advisor, NHTSA	(Authored key law review article on this topic) Automated/autonomous vehicle technology changes how we think about safety and what safety measures need to be adopted (i.e., from crash worthiness to crash avoidance). Past lessons/concepts still useful as a starting point. Past: objective performance test – measure vehicle response from conditions (e.g., described frontal crash test) and used as basis to set performance standards. New Factors: 1. Potentially infinite variables (ball might = child in road) 2. Increased role of software (complex steps) Dealt with similar issues before: Variables – crash testing can occur in various configurations (e.g., number, speed, direction). Came up with set of tests to encapsulate performance and conclude vehicle performance safety in variety of ways that happen in real world. (ex. of specificity of variables: shoes size effect injury in testing) Software – electronic stability control testing will test software's ability to do what intended to do. Following method of crash testing, exposing vehicle to conditions where software needs to intervene or it will fail test. Advantage: testing software ability to sense, think process and effectuate decision. (Hypo: set performance criteria, choose variables, and evaluate software intervention and decision). Conclusion – reminder that yes, this is a new issue but remembering the past tools, methods, and concepts that we've developed and researched helps get us to a point where we are confident in getting to a way to evaluate safety performance for new technologies.
Bernard Soriano, Deputy Director, CA DMV (State)	California is the midst of developing regulatory rules for operational use of autonomous vehicles. Co-sponsor of project to develop regulations with Stephanie Morgan. Two years ago CA legislature passed law for DMV to allow autonomous vehicle on roadways. Definition, prescriptive items of what to include don't match industry (SAE 1 – 5, NHTSA 1 – 4). This is new territory for the State DOT and DMV to develop regulations, so they enlisted help of sister agencies in CA state government and formed steering committee to provide guidance. CA regulations are a potential model for national regulations and NHTSA has regulatory development expertise

	around safety components of autos. Two sets of regulations: 1) Manufacturer testing on public roads; 2) Operations/deployment of autos. Testing regulations are done, approved, and effective as of September 16, 2014, legal to test autonomous vehicles on public roadways. Operational regulations need to be approved. Following public comment period, CA DMW has to formally respond to comments to determine whether changes are necessary (held several public workshops including via social media – Reddit, LinkedIn, etc.) Manufacturer has to apply and notify state whether vehicle can be operated without a human (NHTSA level 4); has to notify CA legislature if vehicle is about to be released; and state has 180 day contemplation period before approval. The state asks manufacturers to disclose areas of operation the vehicle is designed for (urban, freeway, rural – definitions tied to CA code) and asks manufacturer for some data collected during testing period (e.g., VMT, hours in testing, report of any unintended disengagement of autonomous technology). Manufacturer is required to certify they have a fully functional safety plan/program and that sensor data are recorded 30 seconds prior to collision. Manufacturer also needs to provide notice to vehicle operator and occupants that data are being collected by the vehicle. Big sticking point: definition of "the operator of the vehicle" -- need to refine more regulations and vehicle code. A brief overview of the operational steps for testing is as follows: 1. Operator will notify DOT that it will be releasing AV product. Legislature will have 180 days to consider before deployment. 2. Operator must define conditions AV is designed for, i.e., what kind of roads. 3. Operator must provide data from testing, including VMT in tests. 4. Report must include instances of unintended change in acceleration. 5. Manufacturer must certify safety program, with certain required elements. 6. Data from sensors must be collected 30 seconds prior to collision. 7. Device to record sensor data must be separate from EDR. 8. Disclosure to consumers about what data are being collected about them, occupants and operators. Definition of operator is key; the drafters wrestled with that. Vehicle code has a definition that will need to be refined. Other refinements will be needed, e.g., for distracted driving. 9. Special driver's license will not be required. 10. $5 Million bond will be required. 11. Special license plate will be required.

Maxime Flament, ERTICO	Flament provided a European Union (EU) perspective about networking in automation. He described issues being addressed: deployment paths, regulatory issues, testing, connectivity, benefits, digital infrastructure, cybersecurity, human factors, decision and control. Vehicle and Road Automation (VRA) network: commissioned by EU to keep an eye on what's going on in Europe and internationally. The mechanism to support EU Commission specific topics on VRA; contribute to the United States, Japan, EU, and look different aspects of deployment needs. Notable EU activities on legal issues: 1. General Remarks SAE level of automation adopted broadly in EU Confusion of terms – different ways thinking about automation; autonomous (no external help) vs. automated (opposite of manual); automated vehicles connected vs connected automated vehicles; glossary of term Response 4 for highly automated driving, another glossary from railway standards. 2. Legal Framework International Conventions – cross boarder coherence European Level National (translated into each country's language) Vienna Convention amendment by UN Working Group on Road Traffic Safety (ok if can be switched off or overridden by driver) Q: does driver need to be in car? (Netherlands) –answers come from public authorities and traffic code (idea of driver as hardware that drives the software). Netherlands regulatory initiatives – frontline; Sweden preparing for testing in 2016/17 partially or fully automated vehicles; Finland – looking at urban areas max 40 km/h Testing and Validation – 74 tests for manufacturers defined standards have to be tested for can't do it by yourself France, Germany – EU national initiative addressing legal issues; also reviewing all laws, new legal framework 3. Liability Use of data recorder accepted by OEM as necessary to deploy auto vehicles. Tort liability – equipped with adaptive and learning ability implying a degree of unpredictability in the robot's behavior. What happens if damage is not derived from a defect of the robot, but from its behavior? Is a parental model what is needed? Do we need to equip cognitive robots with an ethical code of conduct?
Matthew Schwall, Exponent (Certification and Testing)	Matthew Schwall works for a firm that develops and conducts vehicle tests. He discussed whether AV ratings tests would serve as accelerators or brakes. He noted that tests are distinct from regulations. They serve to inform consumers and encourage manufacturers to make safer vehicles. While cars are improved based on the tests, some part of the improvement may come from "teaching to the test." When a new test is introduced, in many cases the ratings quickly move from "mostly poor to 100% good." For example, a new test introduced in 2012 shows distinguishing ratings; later, it likely will not. Manufacturers tune the system to perform better.

	Tests encourage uniformity. Liability is less likely to be found with uniformity, but the question was raised about whether it is good to have a uniform fleet in a climate of innovation? The question was also raised of how we know that the test is testing the right thing? Tests might mislead the consumer. In some instances, designing to a test may not encourage safety. E.g., oversensitive braking may pass test, but not be best for safety. How ratings would apply to automated vehicles – some have developed test for autonomous breaking systems. Manufacturer doing better now that they have target to shoot toward (April 2014 vs. July 2014 test results). Ratings tests can be a double edged source: can be of use encourage uniformity in performance Positives: technological benefit acting more uniformly (perhaps safer); liability from manufacturer perspective Drawbacks of having rating tests: yes inform consumer but might mislead; designing to test does not necessarily encourage safety; is a homogenous fleet what we want as researchers Give thought about when rating tests should be introduced; the functionalities tested, and how.
Timothy Mattson, Bowman & Brooke (Liability)	Products liability defense side attorney for OEM in litigation; real world knowledge of liability issues. How product liability may interact with autonomous vehicle system Confusion about product liability law vs. regulatory law vs. driving laws Product liability law already exists in every state Essential aspect of defectiveness Some under impression that product liability might change to accommodate autonomous vehicles. While possible, he predicts no chance that is going to happen and the law doesn't need to change if manufactures do the right thing and approach design, data, and document retention in a way that anticipates possible litigation Two categories: risk-benefit test and consumer expectation test or both Risk-benefit Test – Plaintiff wins unless Defendant proves benefit of design outweighs risk of design (burden on defendant – CA different than most states have plaintiff prove risk outweigh benefit; feasible at time product manufacture (not time of design). Autonomous vehicles- nature of algorithm development will lend itself to updates being pushed by manufacturers who should plan to push out improvements all the time on systems if they want to avoid liability issues. Consumer Expectation – Plaintiff wins if they prove product did not perform safely as reasonably expected when used or misused in intended or reasonably foreseeable way; will be largely applied to autonomous vehicles, most courts will find jurors can predict what ordinary consumer can expect; jury will expect autonomous vehicles to do at least as good as human driver Chris Urmson Google presentation would be "gold" for defending systems; message: be gathering data and preserving data; worst thing to do is throw away test data as gathering

	Preemption as potential for protecting manufacturers from lawsuits (i.e., Food and Drug Administration drug warnings = exempt from product liability action) Notice heightened concern of legal issues – liability with big red bar; need to be concerned but don't let paranoia cripple or shape what doing and where technology goes should be driven by needs of market and what makes a good product

Summary of Panel Discussion

- Infotainment category – no one knows what it means. It seems to be imprecise notion. By definition a driver distraction?
  - Need to consider as accelerator if thought of as most broad definition possible, brings some benefit to use.
  - Causes regulatory heartburn from driver distractions standpoint
  - What goal does it have? Info for driver to keep vehicle safe or not necessary for safe driving of vehicle? Difficult to answer.
  - Used because commonly used in RAND study – some thought would accelerate deployment also heard key concerns about navigation systems
- Difficult to develop technology of autonomous vehicle, if so beneficial wouldn't we want to change product liability law to be beneficial for society? If state wants to encourage deployment, what would be barriers for them to soften shared liability regime?
  - Could do that, tort reform, medical malpractice caps on non-economic practice could extend to autonomous vehicles. Reasons why won't happen: 1. Every product will come in and want the same deal; 2. A political problem (trial lawyers, plaintiff attorneys); AV argument not powerful enough to get special exemption under law
  - Other exemptions joint and severable liability – don't see getting special treatment
  - After market automated system placed on system manufacturer vs. vehicle manufacturer
- How are Certification and Testing tests different from NHTSA?
  - Don't design test, just recreate accident driven by needs of client
- Recalls as related to product liability of automated vehicle manufactures.
  - During recall manufacturer acknowledges problem that needs to be addressed. Talking about allegation of design defect.
- Is California conscious of being model in development of regulations? Discussion about federal role vs state for some part?
  - Talk about that quite a bit. Interested in learning about intentions in terms of their regulations moving forward; prescriptive vs. performance based; some aspects could be extended but that hasn't been discussed yet
  - Coordinated effort between DMV to coordinate with each other; vehicles will be crossing state lines; active working group through national vehicle association – what are items you need to consider, insurance driver qualifications
- How will NHTSA do cost-benefit analysis for rule on something not out yet?
  - Depends on what you are doing with regulation (Functionality? Equipment? Performance function?)
  - Not having on-road data is a challenge; use of lab tests

IV. Key Results

The final portion of the Day 2 session was spent reviewing key messages from the first three panels and translating that into key findings and recommendations for next steps

- Regarding the "lead mode" question, the group recognized the attention platooning has received, and several factors that may lead to early deployment of this technology, such as potential cost savings and a cost-conscious industry that will push for its deployment. However, in thinking about what is likely to be the first higher-level deployment, low speed passenger shuttles appeared most likely.
- The group had some difficulty interpreting the role of infotainment. While several technologies appeared useful, and could develop attractive revenue streams, the group found as many potential legal problems in the current environment (e.g., distracted driving) as functional solutions.
- Data and liability continued to be key areas of concern. While some idea of what data are protected, and what data should be protected, is becoming clearer, there was a keen awareness that a significant data breach could significantly set back deployment.
- Similarly, liability appears to be acting as a potential brake on deployment, with concerns about lack of clarity in terms of the ultimate responsibilities of OEM's vs. after-market manufacturers, and lack of consensus regarding true "fail-safe" options (including skepticism of humans serving in that role).
- However, there was strong support for the creation of uniform laws, regulations or standards that would set a common "floor" for these technologies, with the belief that such a standard would accelerate deployment by creating a level playing field of common basic requirements and encouraging competition to exceed these standards at lowest cost. This enthusiasm, however, was tempered by the recognition that overly strict regulations could stifle further development and deployment.

V. Next Steps

- Given the advantages seen for developing a common regulatory "floor," steps toward considering and creating such measures should be undertaken
- The lack of agreement regarding the potential liability risks and protections indicates a need for further dialogue among the legal and technology communities towards developing a common understanding about how risk and uncertainty in this area can be managed most effectively.

Appendix E: The State and Future Direction of Automated-Vehicle Human Factors Proceedings

Session and Proceedings Organizers: Janet Creaser (University of Minnesota), Greg Fitch (Virginia Tech Transportation Institute), Michael Manser (Texas A&M Transportation Institute), Chris Schwarz (University of Iowa)

I. Session Focus and Goals

The success of automated vehicles ultimately hinges on how well they meet their users' needs. The study and application of human factors throughout the automated-vehicle design cycle can yield a safe, useful, and reliable technology that does what its users want. The goal of this breakout session was to present the state of automated-vehicle human factors research and how it is being applied in the development of automated vehicles. The breakout session featured presentations from scientists and engineers leading the investigation of automated-vehicle human factors. Discussion panels provided attendees with a broad perspective of the issues and how they relate to one another. Attendees walked away with a better understanding of recent developments and what the immediate human factors research gaps are.

II. Day 1 Presentations and Discussion

Prof. Dr. Lutz Eckstein (Institute for Automotive Engineering, RWTH Aachen University), Update on Human-Machine Interface Design in Europe

Prof. Dr. Lutz Eckstein is the director of the Institute for Automotive Engineering at RWTH Aachen University. His previous work experience includes leading the HMI and ergonomics development at BMW, MINI and Rolls-Royce, and working on active safety and driver assistance systems at Mercedes-Benz. Dr. Eckstein has been appointed by the European Commission as one of eight European experts in the field of HMI and is working to revise the European Statement of Principles on in-vehicle information and communication systems. Professor Eckstein and his scientists are currently conducting interdisciplinary research on future road vehicles to help ensure that they significantly increase safety and efficiency but also driving pleasure.

Dr. Lutz Eckstein presented an overview on recent European Projects related to the issue of Human Machine Interaction), e.g. ADAPTIVE and UR:BAN. He described how a generic model to structure the large variety of research questions on automated driving is being proposed through publicly-funded and industry projects. The operational definitions for the levels of automation were compared and contrasted between Europe and the United States. A harmonized definition was then presented. The presentation then outlined the research needs from a European perspective which were identified through discussion within academia and with the automotive industry. The research needs included functional design criteria, controllability of functional limits and failures, as well as user acceptance. Finally a combination of methods was proposed to address the various research questions in a well-structured approach.

Johann Kelsch (German Aerospace Center - DLR), EU-Projects on Automation of Road Transport and the iMobility Forum Working Group: Human Factors Perspective

Johann Kelsch is researcher in the German Aerospace Center at the Institute of Transportation Systems. He has been conducting research in the fields of Human Factors, systems ergonomics and human-machine interaction design for highly automated road transport systems for the past 10 years. He is currently involved in various research projects, including the German H-Mode study, and EU-Projects SPARC, CityMobil 1&2, HAVEit, InteractIVe, D3CoS, HoliDes and AdaptIVe. Mr. Kelsch leads the Human Factors subgroup at the iMobility Forum supported by the EU and ERTICO.

Mr. Kelsch spoke on how the development of highly automated road transport systems is becoming more and more complex. He described how system usability becomes crucial as the number of 'differently intelligent' assistance systems increases inside and outside of highly automated vehicles. He then presented the joint system approach and how it uses knowledge from the research area of Human Factors to identify and design for the human-related problems within a joint system. That is, the joint system approach can help develop new concepts for system solutions. It can also reduce system's complexity, increasing usability, reducing costs, and reducing development time.

Mr. Kelsch described his work on the EU-Projects HAVEit, InteractIVe, D3CoS, AdaptIVe as well as his work in the Human Factors subgroup in the iMobility Forum. This description was related to the components of a joint driver automation system. Several examples were presented regarding the systematic exploration and design of system observability and controllability. The examples included different HMI concepts and strategies, such as the 'safety shield', ambient display, driver decoupling, automation levels and cooperation modes. For some examples, experimental data were presented and discussed.

Jim Foley (Toyota), Collaborative Safety Research Center's Benefit Analysis of Pre-Collision Braking

Dr. Jim Foley is a Senior Principal Engineer in the Collaborative Safety Research Center at the Toyota Technical Center. Dr. Foley has over 30 years of experience in Automotive Human Factors and Automotive Safety. He works with leading universities throughout North America to improve automotive safety. Prior to joining Toyota he was at Noblis in Washington, DC, where he provided human factors expertise to NHTSA and the U.S. DOT. He also worked at the Ford Motor Corporation and Visteon for over 10 years. Dr. Foley received his PhD from Purdue University.

Dr. Foley presented Toyota's development of collision targets for a pre-collision system (PCS) performance evaluation. The presentation mainly described Toyota's procedures for evaluating PCS performance. The insight gave rise to discussion on how advanced systems are evaluated and how human factors experts can help better prepare test scenarios.

Steve Casner (NASA Ames Research Center), What Can Automobile Automation Learn From Aviation Automation?

Dr. Steve Casner is a research psychologist at the NASA Ames Research Center. Dr. Casner obtained his Ph.D. from the Intelligent Systems Program at the University of Pittsburgh. He also obtained a Federal Aviation Administration (FAA) Airline Transport Pilot certificate. Dr. Casner described how the introduction of automation to the automobile is following an approach similar to that used to bring automation to the airline cockpit. He described how technology was used to incrementally automate portions of the flying task, and how pilots were left to perform those portions of the task for which engineers were still working towards automating. A number of problems resulted from this process of gradual takeover of a job once performed entirely by humans. Pilots struggle to maintain awareness of a task with which they are no longer intimately involved and of complex technologies that they do not fully understand. Particularly difficult problems arise when increasingly reliable automated systems reach the limits of their capabilities and require pilots to suddenly exercise manual control and reasoning skills that have slipped away as a result of disuse. Dr. Casner argued that road vehicle automation will face these problems and more. Aviation enjoys a broad system of redundancies and safety nets, and, high in the sky, more time to address any problems that arise.

III. Day 2 Presentations and Discussion

Wendy Ju (Center for Design Research, Stanford University), Mental Models for Automated Driving

Dr. Wendy Ju is the Executive Director for Interaction Design Research in the Center for Design Research at Stanford University. Dr. Ju presented an overview of research activity underway in the Stanford Driving Simulator. Her research is focused on developing a set of psychological principles that will guide driver-vehicle interface design that provides effective, real-time support for drivers of increasingly autonomous vehicles. Her team is using controlled driving simulator experiments, as well as in-simulator design improvisation, experimentation and validation methods to develop guidelines and best practices for designing the human-machine interactions in partially automated vehicles. The factors influencing the driver's mental model are being investigated, which yields insight on what the driver expects the car can and will do in a given situation.

Dr. Ju reported that they have found evidence that people are less trusting and comfortable with partial automation than with full automation, and that periods of full and partial automation negatively impact subsequent manual driving performance. The lab's upcoming work will consist of Wizard of Oz studies that allow experimentation of the messaging and interaction behavior between the driver and vehicle. These experiments will be used to see what factors help the driver form an accurate mental model. This will be the basis for developing a framework for designing vehicles that cultivate trust in the vehicle, comfort in the driving experience, and appropriate levels of situation awareness and safe driving behavior that neither over-relies nor under-utilizes the car's intelligent systems.

Janet Creaser (University of Minnesota) & Bobbie Seppelt (Touchstone Evaluations), Workload and Situation Awareness in Transfer of Control

Janet Creaser is a Research Fellow in the HumanFIRST Lab at the University of Minnesota. Her research focuses on examining driver behavior with respect to normal driving conditions and developing methods and technological interventions to support driver safety and mobility. She obtained her M.S. in experimental psychology at the University of Calgary in 2004, with an emphasis on perception, aging, and applied cognition. She is currently pursuing her Ph.D. in Human Factors and Ergonomics at the University of Minnesota.

Mrs. Creaser spoke on the effects of increasing levels of automation on situation awareness (SA). Her talk described how this area of research has been extensively studied in other domains (e.g., aviation, processing control). The main lesson is that operators often experience losses of SA when required to monitor systems with high levels of automation. For automated driving, research indicates drivers' SA is often improved at lower levels of automation (e.g., adaptive cruise control only) as long as the driver does not engage in distracting secondary tasks. However, recent research on higher-automated driving (e.g., adaptive cruise control and lane keeping assist active) indicates a strong propensity for drivers to direct their attention away from the driving environment and towards secondary tasks. This desire to engage in secondary tasks should be assumed and the design of automated vehicle systems should be prepared to effectively bring the driver back into the loop. The presentation then examined a theory of SA in conjunction with design guidelines that could be used to accomplish this difficult task. The presentation discussed how design guidelines relate to presenting the current system status, the boundaries and limitations of the systems, as well as possible ways to present concurrent driving environment information to convey roadway context to drivers of automated vehicles.

Tsutomu Iwase (Subaru), Next Generation EyeSight and Future Technology

Mr. Tsutomu Iwase is the Project General Manager of Advanced Safety Development within the Subaru Engineering Division at Subaru (Fuji Heavy Industries, Ltd.). He has been in charge of developing Subaru's advanced safety technologies, including the latest EyeSight since 2012. He joined Subaru in 1986 and was responsible for various passive safety projects. He has significantly contributed to Subaru's safety in both passive and active safety areas. He holds Bachelor's degree in Mechanical Engineering from Waseda University, Japan.

Mr. Iwase presented on Subaru's next generation EyeSight technology and future strategy. He described Subaru's approach for ADAS (Advanced Driver Assistance System), and focused on Subaru's stereo camera system called EyeSight. EyeSight has played a critical role in achieving Subaru's goal of "Zero Accidents by Automobiles." This is because Eyesight provides better safety and reduces driver workload in all driving-stages. Mr. Iwase showed how EyeSight recognizes various types of objects (e.g. vehicles, bicyclists, pedestrians) simultaneously for functions including Pre-collision braking, adaptive cruise control, lead vehicle start alert, and lane departure warning. He then went on to explain how Subaru's current EyeSight has a 75% equipment rate since its introduction in 2010 in Japan. Subaru's Next Generation EyeSight combines a complementary metal-oxide semiconductor (CMOS) stereo camera (which has color recognition and a wider detection range) with additional steering control to provide new and improved safety functions. Specific configurations and functions were explained, including Active Lane Keep System, Improved Pre-Collision Braking System/ACC, and Pre-Collision Reserve Throttle Management. Mr. Iwase concluded by describing how EyeSight continues to be Subaru's core technology for future strategy, and how it will be used by Subaru's automated vehicles in the future.

Greg Fitch (Virginia Tech Transportation Institute) and Chris Schwarz (University of Iowa), Research Priorities, Gap Closure, and Automation Scenarios in Relation to 2013 Research Needs' Statements (Group Activity)

Dr. Gregory Fitch is a Research Scientist at Virginia Tech's Transportation Institute and leads the User Experience Group in the Center for Automated Vehicle Systems. He received his Ph.D. from the Grado Department of Industrial and Systems Engineering at Virginia Tech in 2009.

Dr. Chris Schwarz is the Simulation Software Team Leader at the National Advanced Driving Simulator University of Iowa. He received his Ph.D. in E.C.E. from the University of Iowa in 1998. Dr. Fitch and Schwarz led a discussion on human factors research priorities.

Dr. Fitch began the discussion by reviewing the role of the Human Factors breakout group and the 2013 research need statements. A primary debate within the breakout group over the course of the symposium centered on whether it was appropriate to develop partially automated vehicles, mainly given drivers' likelihood of misusing the automation, limitations in maintaining SA, and limitations in re-engaging in the driving task under a short period of time. At the same time, data from the NTSB were presented that showed the steady reduction in aviation accident rates with each successive generation of aviation automation. This gave way to a discussion of what the overall transportation system performance might be with the introduction of prevalence of partial automation. Discussion then moved on to debate the definitions of the levels of automation. The point was made that the definitions serve a discussion of vehicle automation. Whether partial automation is appropriate or not, the definitions enable research to be scoped to generate human performance data. It is highly important that policy and design decisions be made based on human performance data. Examples of how human factors has been successfully incorporated into iterative user-centered design of driver assistance systems were given. The process consisting of identifying driver errors with prototype systems early in the design cycle and using the data to inform future design iterations. This discussion transitioned into a discussion on what human performance data are needed now. There is an urgent need to develop a deeper understanding of: 1) driver misuse and abuse of automation, 2) the transfer of control between vehicle users and automation, 3) how to effectively convey automated vehicle status, and 4) how to make automation adjustable to the user's preferences.

Dr. Schwarz presented several videos that exemplified automated vehicle use cases. A discussion ensued on how drivers/users might handle automation in specific scenarios, including the transfer of control on a highway and a highly automated intersection. Test scenarios recommended based on a concept of operations document prepared by the Virginia Tech Transportation Institute were presented. The test scenarios were identified based on a review of typical crash observed in naturalistic driving databases as well as from domain expert insight.

IV. Key Results

- Operator interaction with partial automation is documented within aviation research. However, there are many significant differences between aviation and surface transportation, notably:
    - Pilots are highly-trained, and hand-selected, to operate a plane, where light vehicle drivers are relatively untrained and present more variability in capabilities and driving skills.
- There is substantial space between planes in the sky leaving pilots numerous seconds to respond to unforeseen events, where light vehicles are much closer to each other leaving drivers milliseconds to respond to unforeseen events.

- Driver misuse (unintentional improper use) and abuse (intentional improper use) of commercially-available partial vehicle automation is becoming a growing concern. There have been numerous videos posted online of drivers circumventing precautions vehicle designers have taken to ensure drivers remain in the loop with the driving task. Examples include drivers fastening soda cans to the steering wheel to trick the automation into sensing that a hand is on the steering wheel. Research is needed to investigate real-world driver behavior with partial automation in order to characterize the different ways drivers misuse and abuse the technology.
- The effective and safe transfer of control from automated vehicles to drivers remains a top research priority. Simulator research has been performed since the previous workshop to investigate the human factors issues, but remains to be fully understood. The alert modality, timing of the alert, progression of alerts, and driver state need to be considered.

V. Next Steps

The next steps for this breakout group include the following:

- Conduct a naturalistic driving study of driver use of commercially-available partial vehicle automation.
- Investigate the timing and modality of take-over-requests as well as the role of the environment and driver state in effectively regaining control of the vehicle when the automation becomes unable to operate in an upcoming scenario.
- Develop a comprehensive understanding of drivers' mental models of automated vehicle operation and what information is needed from the interface to inform/update these mental models.
- Investigate the role of driver monitoring to develop an equal understanding of the drivers' state as the state of the environment.
- Involvement of human factors research early on, as well as throughout, the design of automated vehicles to ensure user interaction with the automation is well understood and supported.
- Provide an update at the 2015 TRB Annual Meeting workshop.

Appendix F: Near-Term Connected/Automated Technology Deployment Opportunities Proceedings

Session and Proceedings Organizers: Robert Ferlis (FHWA), Ginger Goodin (Texas A&M Transportation Institute), Thomas Bamonte (North Texas Tollway Authority) , Ram Kandarpa (Booz Allen Hamilton), Suzanne Murtha (OmniAir Consortium), Joseph Peters (FHWA), Shelley Row (Shelley Row Associates), Chris Armstrong (Leidos), Anna Giragosian (Leidos)

I. Session Focus and Goals

Automated vehicles by themselves can deliver many but not all of the benefits possible through automation. Automation assisted by connection to other vehicles and the infrastructure and by roadway features such as managed lanes may be needed to unlock the full potential inherent in automation. Connected automation systems that include a role for an engaged driver can provide many benefits in the nearer-term, as part of the decades-long transition to a fully automated highway system. This breakout session focused on ways to advance the development of an automated and connected highway system that will support increasingly capable vehicles and deliver safety and highway system operational benefits both in advance of and in addition to the benefits associated with fully automated vehicles.

It is important to identify these near-term opportunities in order to engage state and local highway transportation authorities and the automotive community. Public highway authorities for the most part tend to view fully automated vehicles as a consumer product that will deliver relatively few benefits for highway operations. Can incremental technologies that deliver near-term highway capacity and operational benefits help turn highway authorities from bystanders to advocates for an automated and connected highway system? If so, agencies can create a supportive technical and political environment for, among other things, the licensing and deployment of automated vehicles developed by the automotive community.

This breakout session focused on near-term deployment opportunities and concluded with a discussion of how to integrate these near-term opportunities with ongoing efforts to deliver a fully-automated highway system.

II. Day 1 Presentations and Discussion

Opening Remarks

- **1:45pm – 3:30pm: Technology Discussion –** Robert (Bob) Ferlis, Federal Highway Administration
    - Mr. Ferlis welcomed everyone to Breakout Session #7 and introduced the topic of Near-Term Connected/Automated Technology Deployment. He explained the format for the session and introduced the four key topic areas for discussion across the two days: Technology, Innovators, Early Deployment, and a final session for discussion of the Transition to Long-Term Deployment.
    - Mr. Ferlis moderated the Technology topic by introducing each panelist and proposing the two questions which were to be the focus of the discussion:
        - What are near-term connected vehicle and infrastructure technologies that can deliver safety and operational benefits over the next several decades?
        - Can we advance the capacity enhancement, traffic flow and safety benefits before many or all vehicles are fully automated?

- **3:45pm – 5:30pm: Innovators Discussion** – Ginger Goodin, Director, Transportation Policy Research Center, Texas A&M Transportation Institute
    - Ms. Goodin welcomed everyone to the discussion on Innovators. She introduced each panelist and explained how the Innovators topic fit within the context of the breakout session.
    - Ginger then facilitated questions after each presentation and at the conclusion of all presentations focusing on answer the following questions:
        - Who is best situated to champion the early deployment of automated vehicle/highway technologies?
        - Can we resolve the current conundrum where state and local highway authorities are not especially motivated to support automated vehicle technology and vehicle developers view such infrastructure providers/operators as unreliable and unhelpful potential partners?

Panel/Presenter Summaries

- The Day 1 Proceedings were focused on two separate topics: **Technology** and **Innovators** in the context of the overall breakout session subject.
- First, **Bob Ferlis** moderated the **Technology** session where each panelist gave a short presentation on near-term deployment opportunities from their perspective followed by a brief question and answer session related to the presentation.
- **Ginger Goodin** followed by moderating the **Innovators** session where each panelist gave a short presentation on their perspective and answered follow-up questions on the session topic.

Speakers	**Summary of Presentation, Question and Answer Session**
Technology Session	
Dr. Steven Shladover, Program Manager, Mobility at the California PATH Program of the Institute of Transportation Studies at University of California Berkeley *Near-Term Automation Opportunities*	• Presented a snapshot of when society may expect market availability across the various steps of implementation for each level of automation o Full automation (level 5) everywhere is far away (2040s) o Limited access highways provide the greatest opportunity and urban streets are the greatest challenge for the higher levels of automation • Suggested that I2V connectivity is a key step towards levels 1 and 2 automation benefits where the following may provide the greatest opportunity: o Limited access highways o Signalized arterial infrastructure to vehicle CACC • Highlighted actions needed to realize benefits: o Connected vehicle market penetration o Concentrate equipped vehicles near each other to increase "virtual" local market penetration (dedicated lanes)
Kevin Dopart, Program Manager, Vehicle Safety & Automation, U.S. DOT ITS Joint Program Office *Connected Vehicle Path to Deployment*	• Presented a timeline on the U.S. DOT connected vehicle path to deployment including current and upcoming activities • Provided examples of current level 1 research activities o Driver advisory and warning applications using limited control functionality (CACC, SPD-HARM, Q-WARN) o Eco-Signal Operations • Summarized deployment challenges identified – aftermarket devices,

	spectrum sharing demands, communications congestion potential, and other road users
Michael Maile, Autonomous Driving North America, Mercedes-Benz Research & Development North America, Inc. *Vehicle-Infrastructure Communications-based Safety Applications*	• Presented Mercedes Benz history on developing vehicle communications applications – began V2V in 2002 and V2I in 2004 – and presented status of current research • Suggested that V2I scales a lot better than V2V providing the example that a location can provide significant benefit to a local area before fleet wide implementation could become a reality • Suggested that vehicle connectivity is not going to enable autonomous driving but is going to make it better • Presented plans to refine and select applications for implementation using DSRC test beds
Christian Schumacher, Advanced Driver Assistance Systems, Continental Automotive Systems, N.A. *Near-Term Connected/Automated Technology*	• Presented on the various sources of information which drive system and product requirements for Continental products o Sensors provide info on immediate surrounding environment o Backend data provide information beyond what is "seen" by the vehicle o Data fusion brings all info together • Suggested that connectivity, backend data, and map data are key to further enabling automation • Suggested that automated capabilities have to handle any and all driving situations independent of connectivity, but that connectivity can better enable safety in many situations
Greg Larson, Chief of the Office of Traffic Operations Research, Division of Research, Innovation, and System Information *Near-Term Opportunities for Automated Vehicles*	• Suggested that in an ideal world, connectivity would have come first and automation would have been built on top of it • Highlighted National Automated Highway System Consortium work which was completed in 1998 • Highlighted previous and current Caltrans and University of California system automation research – passenger car and bus demos, truck platooning, and current Oregon vehicle assist and automation operating today in revenue service • Summarized operational considerations – determine optimal balance of intelligence between vehicles and infrastructure; emphasize heavy vehicle applications as early adopters; and determining whether to start with mixed traffic or exclusive right-of-way
Innovators Session	
Bob Frey, Planning Director, Tampa-Hillsborough Expressway Authority *Driving Innovation and Opportunity: Tampa Bay is ready to connect to*	• Presented how Tampa-Hillsborough Expressway Authority has become a leader in pursuing automated technologies as a small, local agency. Reasons include history of innovation (e.g., first reversible elevated lanes in world); supportive leadership (state legislators, Executive Director, Board of Directors), and Innovative partners • Provided an overview of steps Tampa-Hillsborough Expressway Authority has taken to become actively involved in automated vehicles o Co-hosted AV/CV Summit with Florida Department of Transportation o Signed U.S. DOT Connected Vehicle Test Bed Memorandum of

Autonomous Vehicle Technology	Understanding o Participating in Florida Department of Transportation's Statewide initiative working group o Actively marketing Expressway as test bed; testers including Volkswagen have signed up • Stressed the importance of smaller agencies getting involved early to determine what the impacts of technology will be on them and to voice any concerns
Casey Emoto, Deputy Director, Product Development, Santa Clara Valley Transportation Authority *Finding Champions through Collaboration*	• Described how champions were fostered for the Bay Area Express Lanes through collaborative initiatives such as regional and countywide transportation plans and other local, regional, state, and federal efforts • Suggested that linkages between AVs/CVs and managed lanes include that both: allow for more roadway throughput without adding more lanes; are focused on managing the roadway facility and reducing travel times; and involve new technology that should be tested and deployed in market niches rather than fully deployed all at once • Discussed that in order to encourage institutional collaboration, agencies need to be presented with a problem that AVs/CVs can help solve
Chuck Fuhs, Consultant *A P3 Perspective*	• Suggested that public-private partnerships (P3) can be influenced by a variety of private and public stakeholders and all perspectives should be considered • Highlighted the perspectives of P3 partners, which include the need for a return on investment, balancing risks with return, and a potential resistance to innovate (at least initially) • Suggested that applying P3s to automated vehicle deployments will require answering questions concerning what the impacts will be on traffic flow, LOS, safety, and reliability, which will in turn require proving the benefits through tests and demonstrations • Discussed the need for ongoing outreach and education to sell consumers on things that may not seem to be in their immediate interest (e.g., speed harmonization requires drivers to go slower instead of faster)
James Dreisbach-Towle, Principal Technology Program Manager, San Diego Association of Governments (SANDAG) *San Diego Motivation*	• Stressed that private industry will need to champion AVs, as they will be motivated by the business proposition of AVs and will be able to progress more quickly • Suggested that the role of public agencies will be to support private sector innovation by providing open operational data, certification-ready roads, hyper-local infrastructure, and by raising awareness among local stakeholders • Emphasized the need to establish trust among stakeholders, decision-makers, and the public through ongoing outreach and education o Regional government can assist with establishing trust by fostering decision-maker engagement, incorporating AVs into long-range planning, standardizing network assets, and raising general awareness
Collin Castle, Connected Vehicle Technical Manager,	• Suggested that MDOT's role is to support what AVs can do; ongoing efforts include the Toward Zero Deaths Initiative, Michigan Senate bills on Automated Vehicle Testing (SB169) and Companion Liability (SB663),

Michigan Department of Transportation (MDOT) *Michigan DOT AV Activities*	and MDOT support of AV testing and research • Described CV infrastructure deployments in Michigan, including the U.S. DOT Southeast Michigan Test Bed, the U.S. DOT Safety Pilot Model Deployment, and the MDOT 500 Roadside Unit Deployment Initiative • Provided an overview of the University of Michigan Mobility Transformation Facility , which will include physical and IT infrastructure and will provide public and private access for infrastructure testing to support AVs • Discussed that maintaining AV/CV infrastructure may be a challenge at the local level, where there are multiple jurisdictions and revenues are down

Summary of Panel Discussion

- **Third party automation products.** Market-available automation technology has primarily been delivered by OEMs to date. However, there are now third party vendors selling aftermarket systems which promise automation functionality with retrofit equipment taking advantage of OEM actuation and activation technology
    - Are there legal and liability concerns from a manufacturer and/or public agency perspective? Should these third party systems be allowed?
    - OEM – Anytime an aftermarket product interacts with actuation or activation systems which are capable of making driving decisions, there is a concern. These aftermarket products likely don't go through the rigorous validation and testing methods performed by OEMs to ensure safety and handle failure. Ultimately, the consumer purchasing the system would be responsible for any liability.

<u>Technology</u>

- **OEM and supplier perspective on connectivity and automation.** Vehicle connectivity is not going to enable autonomous driving but is going to make it better. Automated capabilities have to handle any and all driving situations independent of connectivity, but connectivity can better enable safety in many situations. Connectivity, backend data, and map data are key to further enabling automation.
- **Deployment challenges from U.S. DOT perspective.** Aftermarket devices, spectrum sharing demands, communications congestion potential, and other road users
- Operational considerations from roadway authority perspective (State DOT):
    - Determine optimal balance of intelligence between vehicles and infrastructure;
    - Emphasize heavy vehicle applications as early adopters; and
    - Determining whether to start with mixed traffic or exclusive right-of-way
- **Researcher perspective.** Full automation (level 5) everywhere is far away (2040s). Limited access highways provide the greatest opportunity and urban streets are the greatest challenge for the higher levels of automation. Actions needed to realize benefits: 1) Connected vehicle market penetration and 2) concentrate equipped vehicles near each other to increase "virtual" local market penetration (dedicated lanes)

<u>Innovators</u>

- Agency Involvement
    - Agencies – big and small – should be looking at this technology now to determine what the impacts will be

- As agencies conduct outreach and get involved in stakeholder groups, OEMs will start to reach out
- Agencies need to be prepared to add capacity – staff, funding, technical capabilities

- Public role with private sector innovators
  - Providing access to open operational data
  - Leading education to build trust in technology
- Need to clearly communicate the benefits to achieve public sector buy-in and identification of champions

III. Day 2 Presentations and Discussion

Opening Remarks

- **1:45pm – 3:30pm: Early Deployment Discussion –** Robert (Bob) Ferlis, Federal Highway Administration
  - Mr. Ferlis welcomed the audience to the continuation of Breakout Session #7 and summarized the topic of Near-Term Connected/Automated Technology Deployment. He explained the format for the session and reminded the audience of the four key topic areas for discussion across the two days: Technology, Innovators, Early Deployment, and a final session for discussion the Transition to Long-Term Deployment.
  - Mr. Ferlis moderated the Early Deployment topic by introducing each panelist and proposing the two questions which were to be the focus of the discussion:
    - Is there a value proposition that private sector technology integrators can offer to highway authorities and vehicle OEMs to facilitate early deployment of automated vehicle/highway technologies?
    - Are there particular highway configurations, such as managed lanes, that offer better near-term deployment opportunities?
    - What are the value propositions for public and private interests?
- 3:45pm – 5:30pm: Transition to Long-Term Deployment Discussion – Chris Armstrong, Transportation Engineer, Leidos
  - Mr. Armstrong welcomed everyone to the late afternoon discussion on Transition to Long-Term Deployment. He explained the purpose of this final topic area and the need to summarize the discussion from all four session topic areas to report overall breakout session findings and conclusions to be presented on Thursday.
  - Chris then facilitated a discussion focused on answering the following questions related to Transition to Long-Term Deployment:
    - What do we now think the promising near-term (next ten years) deployment opportunities will be? What changes in technologies (automation and other), markets, economy, or society can we expect over the long-term (up to 30 years) horizon that must be considered?
    - How can we best transition from early deployment to long-term deployment?
  - This session will be used to create a summary of breakout session findings and conclusions for reporting on Thursday.

Panel/Presenter Summaries (Complete one row for each panel/speaker)

- The Day 2 Proceedings were focused on two separate topics: **Early Deployment** and **Transition to Long-Term Deployment** in the context of the overall breakout session subject.

- First, **Bob Ferlis** moderated the **Early Deployment** session where each panelist gave a short presentation on near-term deployment opportunities from their perspective followed by a brief question and answer session related to the presentation.
- **Chris Armstrong** moderated a discussion with the audience on the topic of **Long-Term Deployment** and facilitated a discussion to create a summary of the overall breakout session findings and conclusions for reporting on Thursday.

<table>
<tr><th>Speakers</th><th>Summary of Presentation, Question and Answer Session</th></tr>
<tr><td colspan="2">Early Deployment Session</td></tr>
<tr><td>Thomas Bamonte, General Counsel, North Texas Tollway Authority</td><td>Connected/Automated Vehicle Early Deployment
• Proposed that it will require an investment in infrastructure to unlock the true value of AV/CV technology
• Suggested that opportunities for AV/CV partnerships between highway authorities and the private sector include:
o Building safety tools into the CV platform
o Making roadways machine readable
o Data harvesting/operational response techniques
o Improvements through electronics, not concrete
o P3s driven by savings vs. a traditional business model
• Described a possible “grand bargain” between the private sector and highway authorities in which the private sector delivers system capacity improvements through platooning capabilities in vehicles, and highway authorities advance the potential for innovation through investment and legislation
• Suggested that existing examples of P3s involving infrastructure could be used as models for AV/CV P3s, though different players may be involved</td></tr>
<tr><td>Steve Lockwood, Parsons Brinkerhoff</td><td>• Outlined transactions among various players in the AV/CV arena, including private systems/technology providers, customers/users, public sector and private third parties
• Described five major value propositions for AV/CV:
o Deliver minimum early payoffs to auto users in safety ASAP
o Deliver maximum early payoffs in efficiency for freight and transit fleet operators ASAP
o Possible program structure to assist policy decision-making to support realistic/optimum paths toward AV/CV options regarding key payoffs/trade-offs in safety and efficiency
o Supplement state DOT’s capabilities in order to accelerate V2I infrastructure deployment
o Provide policymakers/public with public benefit rationale to support progress in AV/CV
• Discussed what the benefits of level 1 and 2 technologies were expected to be, and stressed the need for the public sector to provide settings where these benefits can be tested
o Level 1 technologies are expected to provide system-level/public sector benefits, but consumers may not be motivated to pay extra for these technologies</td></tr>
<tr><td>Suzanne Murtha, OmniAir and Atkins</td><td>Certification of Autonomous and Connected Vehicles: Ensuring Safety and Reliability
• Presented purpose and activities of OmniAir Consortium</td></tr>
</table>

	• Suggested that autonomous vehicle certification can enable near-term deployment by ensuring confidence in some level of safety and allowing more control from industry, though interest in legislating and certifying varies by state • Explained how connected autonomy can provide life saving benefits through high speed connectivity to near and far field vehicles and to infrastructure • Discussed whether drivers will need to be certified to operate AVs, and if so whose responsibility it will be to ensure their certification
Doug Davenport, Prospect Silicon Valley (501(c)3)	**Next Generation of Technology Solutions: For Cities Around the World** • Highlighted Prospect Silicon Valley activities to accelerate path to deployment by offering tech companies: access to critical infrastructure, use of demonstration/commercialization center with support (labs, high bay garages, fully connected intersection, etc.), and unique multi-stakeholder programs – concept to customer • Presented North San Jose Innovation Center – proving ground for advanced transportation systems • Highlighted partnerships that are supporting the deployments and field tests planned
Chris Schreiner, Strategy Analytics	**Near-Term Connected/Automated Technology Deployment Opportunities: Consumer Perceptions and Impact** • **Consumer Interest**: compared to other ADAS, consumers express far less interest in autonomous driving; more than 1/3 not at all interested in autonomous highway driving; 40 percent not at all interested in full autonomous driving; and the word "autonomous" has negative connotations and drives interest down o This is great for somebody else – limited mobility, distracted, etc. • **Willingness to Pay**: not very strong for autonomous features; only 10 percent willing to consider paying $10k extra for fully autonomous; even at extremely low price point ($2k), only 27 percent would consider paying • **Consumer Apprehension**: issues with trust; issues with control • **Level 3 Autonomous Vehicle**: robustness of autonomous solution affects handover of control; non-fully autonomous requires conservative restrictions on drivers; restrictions negatively impact usefulness, consumer interest, and acceptance

Summary of Day 2 Results/Session wrap-up

<u>Early Deployment</u>

- Roadway authority perspective:
  - Google/OEMs deliver capacity improvements via <u>platooning</u>
  - Highway authorities advance system technologies via <u>investment/legislation/policy</u>
- Discussion of value propositions for automated vehicles
  - Deliver maximum early payoffs to auto users in <u>safety</u>
  - Deliver maximum early payoffs to freight users and transit fleet in <u>efficiency</u>

- Provide program structure to assist Policy decision-making
- State DOTs increase capabilities in order to accelerate V2I infrastructure deployment
- Provide policy makers/public with public benefit rationale to support for full automation

- Role of public sector
  - Establish facilities where we can learn more about the benefits for level 2 and 3 automation
  - Could support certified vehicles access to automated technology lanes – could serve as an example for full market penetration scenario
  - Multi-state partnerships could provide in roadway facilities
- Be aware of consumer opinion and sensitivity to autonomous vehicles
  - Currently, consumers are wary of autonomous features but may not be fully informed of the potential benefits
  - Willingness to pay is low and there are control/trust issues. However, safety does sell
  - Level 5 is the current consumer level of understanding and expectation

Transition to Long-Term Deployment

- What do we now think the promising near-term (next ten years) deployment opportunities will be?
  - Level 1, 2 applications for managed lanes
  - Small scale community-based environments (retirement, campus, residential) offer opportunity for high level automation deployment (level 4 and 5)
  - Public transportation and/or freight offer additional opportunity – specialty truck lanes
  - Automation could enable greater safety and reliability for freight operations (including LCVs) on non-typical facilities – federal guidance could accelerate
  - Automation for entrance/exit at freight facilities
- What changes in technologies (automation and other), markets, economy, or society can we expect over the long-term (up to 30 years) horizon that must be considered?
  - Infrastructure investment – public sector financial capacity
  - Electrification
  - Shared use – personal, commercial
  - Vehicle ownership sociology
  - Liability and insurance framework for automation
  - International standards or guidelines
  - Security / Privacy / Information Sharing
  - Stock versus Aftermarket Retrofit
- How can we best transition from early deployment to long-term deployment?
  - Accommodate progression of technology but allow for consistency with regard to performance
  - Performance based standards that support inoperability amongst the various levels: Level 3 does not necessarily add benefits in safety, mobility, and environment.
  - Carefully consider the roll out and staging of technologies for each level of automation and within each level – a progression approach to functionality and performance may better support buy-in
  - Rethink performance measures when identifying or highlighting safety, mobility, environmental benefits to incentivize automation
  - Industry standardization to ensure legal support across jurisdictional boundaries
  - Key considerations as we progress through deployment: consumer perception/benefit, liability/risk, operational performance, cost/benefit

IV. Key Results

Technology

- **Researcher perspective.** Actions needed to realize benefits: 1) Connected vehicle market penetration and 2) Focus on near-term level 1 applications – CACC, Speed Harmonization, Eco-Signal Operation
- **OEM and supplier perspective on connectivity and automation.** Vehicle connectivity is not going to enable automated driving but is going to make it better.
- **Deployment challenges from U.S. DOT perspective.** Aftermarket devices, spectrum sharing demands, communications congestion potential, other road users, and security
- **Operational considerations from roadway authority perspective (State DOT).** Determine optimal balance of intelligence between vehicles and infrastructure; Emphasize heavy vehicle applications as early adopters; and Determining whether to start with mixed traffic or exclusive right-of-way

Innovators

- Agency Involvement
  - Agencies – big and small – should be looking at this technology now to determine what the impacts will be
  - Agencies need to be prepared to add capacity – staff, funding, technical capabilities
- Public role with private sector innovators
  - Providing access to open operational data
  - Leading education to build trust in technology
- Identify public sector champions who need to address problems that automation can solve:
  - Addressing congestion
  - Providing additional capacity
  - Improving environmental impact

Early Deployment

- **Roadway authority perspective.** Highway authorities can advance system technologies via investment/legislation/policy
- **Identify value propositions for automated vehicles.** Safety and efficiency payoffs, policy supported by public benefit rationale, and increase in infrastructure capabilities
- **Role of public sector.** 1) Establish facilities where we can learn more about the benefits of automation, and 2) sponsor and invest in early deployments
- Be aware of consumer opinion and sensitivity to autonomous vehicles.
  - Currently, consumers are wary of autonomous features but may not be fully informed of the potential benefits
  - Willingness to pay is low and there are control/trust issues. However, safety can sell
  - Level 5 is the current consumer level of understanding and expectation
  - However, mobility benefits are achieved at level 1 and further levels of automation primarily offer consumer convenience – need to bridge this gap

Transition to Long-Term Deployment

- Promising near-term (next ten years) deployment opportunities:
  - Level 1 applications for managed lanes and tollways
  - Public Transportation and Freight applications

  - Small scale community-based environments (retirement enclave, campus) offer opportunity for higher level automation deployment (level 4 and 5) due to lower risks
- **Changes in technologies, markets, economy, or society that must be considered.** Infrastructure investment, electrification, shared use, vehicle ownership sociology, liability and insurance framework for automation, international standards or guidelines, security / privacy / information sharing, stock versus aftermarket retrofit.
- **How can we best transition from early deployment to long-term deployment?**
  - Accommodate progression of technology but allow for consistency with regard to performance – can standards minimize disruption and preserve benefits?
  - Allow for incremental improvement– e.g., following distances for CACC

V. Next Steps

- Circulate results to the relevant committees for development of research needs
- Encourage a structured dialogue including those committees responsible for developing research road maps
- Consider a task force to facilitate buy-in or identify roadway authorities that are currently participating or willing to participate in automation activities

Appendix G: Personal Vehicle Automation Commercialization Proceedings

Session and Proceedings Organizers: Bob Denaro (Independent Consultant), Alain Kornhauser (Princeton University), John Suh (Hyundai Ventures), Robert Seidl (Transportation Technology Ventures)

I. Session Focus and Goals

Simply stated, the focus of the Personal Vehicle Automation Commercialization breakout session was to investigate how automated vehicles will make money. The session addressed the commercial viability of automation SAE levels 3, 4 and 5 on passenger cars sold by the automotive manufacturers to consumers. Value elements of efficiency, safety and other factors were considered. Particular interest was devoted to new market forces such as the insurance industry's positive or negative disruption from automation and the role of information, services and Infotainment companies who now have a new valuable channel to the consumer.

The goal of the PVAC session was and analysis of the various market forces that may impact the emergence of automated personal vehicles, including some that might not be intuitive in today's automotive industry.

II. Day 1 Presentations and Discussion

Opening Remarks

- Organizers Bob Denaro, Alain Kornhauser, John Suh and Robert Seidl introduced the session by encouraging the attendees to think broadly about the market forces facilitating or impeding the emergence of automated vehicles. An example would be retailers offering rebates for accepting advertising and, in particular, location-based advertising.

Panel/Presenter Summaries

- On the first day of the PVAC breakout session four speakers presented on relevant topics for the session, followed by a panel discussion with the attendees.

Speakers	Summary of Presentation, Question and Answer Session
Kevin See Lux Research	• Co-authored "Capitalizing on the $87B Opportunity in Self-Driving Cars, by Lux Research, April 2014 • The hype for automated cars getting ahead of adoption • The cost of LiDAR, an essential sensor for vehicle automation, is an obstacle to adoption of automated vehicles as batteries are for EVs • Level 2 automation, such as ACC, is available today but expensive • Electronic Stability Control and Automated Braking Systems took 11 and 17 years, respectively, to reach just 5% penetration • Beware of possible setbacks with possible accidents and strong press and consumer reaction • A key business model for adoption of automated vehicles might be shared vehicles • Level 2 self-driving will increase from a small fraction of new cars sold today ‹ about 3% globally ‹ to 57% in 2020, and 92% in 2030. However, by

	that same year only 8% of new cars sold will attain the reasonable capabilities of level 3 autonomy, and no level 4 fully autonomous cars will be available. • At first, level 2 will see the strongest growth in Europe and the U.S., but China will quickly catch up in technology to these early leaders, and after 2020 surpasses all in unit sales of level 2 autonomous cars thanks to its rapidly growing overall auto market. • Software will be the biggest autonomous vehicle value chain winner, with $25 billion in revenues in 2030, a 28% compound annual growth rate. This software will be largely invisible to the driver, operating behind the scenes as machine vision that brings sensor inputs together, and that uses artificial intelligence to determine a safe navigation path through the world. It will be a differentiated and high-stakes field, ripe for new partnerships beyond the conventional automotive value chain. • By 2030 automakers will be able to capture profits of about $9.3 billion from the emergence of autonomous vehicles, making this new technology an alluring proposition. However, level 3 autonomy will be a premium option, opening the door to business model innovation if automakers hope to deploy it beyond some high-end vehicles. • Optical cameras and radar sensors will amount to $8.7 billion and $5.9 billion opportunities in 2020, respectively, thanks to level 2 cars. However, because of the increasing complex processing requirements of level 3 autonomy, in 2030 computers will be biggest hardware opportunity on-board autonomous cars, amounting to a $13 billion opportunity.
Mike Scrudato Munich Reinsurance America	• Automated Vehicles could cause a major disruption in the insurance industry • There are two schools of thought in the insurance industry: o The driver is still responsible and liable o There will be a gradual evolution of liability from the driver to the auto manufacturer • Case number 2 is emerging today. Insurance may come with the car at purchase time • There are key factors outside of the insurance company's control, such as how fast legislation evolves and early court tests of the liability question • As for automated vehicles' impact on premium rates, where many expect such vehicles to be much safer than conventionally operated vehicles, the insurance industry must wait until they have an adequate body of data to evaluate risk
Kyle Vogt Cruise Automation	• Cruise automation is developing an aftermarket solution for vehicle automation. As a commercial company, they are focused on building something that consumers want. The business proposition is targeting the U.S. installed base of 250M cars vs. 15M new car production per year. • They are focused on a solution for automation of highway driving • The key question is, will people pay for it? They are taking pre-orders to assess demand. Initial pricing for the system is $10,000. Cruise

	Automation acknowledges that the price/value proposition needs some work. • Cruise Automation is moving to test drives in 2015 for 50 early adopters. The system works on 2012 or later Audi A4/S4 automobiles and is restricted to California highways. • Startups can move faster than traditional automobile manufacturers by skipping steps in the development process • Besides sales of systems, there is value in the data collected by automated vehicles such as digital map updating
Chris Borroni-Bird Qualcomm	• Monetizing the automated vehicle is the challenge • Safety can be achieved with ADAS. You don't need automation to achieve significant gains in automation. • Commercial vehicles may justify automation through savings in fuel economy • Automated vehicles create productive travel time for the consumer • Land use will be strongly impacted with high penetration of automated vehicles. In a largely shared vehicle environment parking requirements diminish. • Shared vehicles and electrification will have a big impact with automatic rebalancing of vehicles to meet daily demand • Door-to-door mobility for the disabled and aged are significant benefits of widespread automation • Automation and electrification lead to more energy efficient vehicle designs and lower cost • Standardization is essential to successful deployment of automated vehicles • Key technology enablers include advanced, low cost sensors; actuators; human machine interface; digital maps; and DSRC for V2V • Qualcomm is experimenting with DSRC in a mobile phone that can interact with vehicles. It would power up only when out of doors and the user is moving.

Summary of Panel Discussion

Key points made with speakers in the subsequent panel discussion

- The driver of an automated vehicle, if present, will be responsible and liable. The insurance company will pay the other party and then go after other parties. But this will need a court test.
- In an automated vehicle, engaging in distracting activities (like updating your Facebook page) will not be a proximate cause of an accident.
- It is not likely that Telecom companies will be liable for distractions in the vehicle. The car manufacturers and legal authorities will set limits on their use.

- Cruise Automation is working on the cost curve, achieving a 4-hour installation, and getting additional car manufacturer agreements.
- A separate license for operating an automated vehicle is not practical
- Level 3 automation is not feasible because consumers will become complacent with the automated mode and not be ready to regain control
- The legislation and regulation problem is not insignificant
- The strongest business case is urban congestion and travel in cities
- There is a strong value proposition for early partial automation such as parking, traffic jam assist, and highway super cruise control, but full automation will pull along shared vehicle business. There may be no business case for a fully automated personally-owned vehicle.
- The lack of law breakers could decimate municipal budgets
- 90% of organ donor organs today come from auto accidents. Dramatically reduced highway fatalities may impact this segment.

Summary of Day 1 Results

- Preliminary observations on day one identified the struggle between the traditional auto manufacture incremental development and product offering process and, albeit overly hyped in the press, disruptive demonstrations of advanced technology by new players. The path that evolves is important to many related industries and sectors such as insurance, advertising, infotainment, governments, traffic control and others.

III. Day 2 Presentations and Discussion

Opening Remarks

- Alain Kornhauser offered the thought experiment of an insurance company offering coupons to other insurance companies' bad drivers which could be redeemed at a dealership offering automated vehicles. The insurance company addresses a neglected market segment and buys down the cost of the automation "option" on the new vehicle.

Panel/Presenter Summaries

- Day 2 of the PVAC breakout session started with two additional speakers and a short panel discussion with the attendees. Then the attendees broke into sub-breakouts to discuss the business case for vehicle automation in SAE levels 3, 4 and 5, considering the market forces that might accelerate or delay adoption.

Speakers	Summary of Presentation, Question and Answer Session
John Capp, GM	• There is too much hype about automation levels • Be careful about assumptions of fast adoption. Adaptive Cruise Control has been in the market for 15 years and adoption today is still very small • While automation features might be offered in a particular year, strong penetration in the consumer market may take many years. For example, while automated cruise control may happen in 2018, it could be 20 years

	later until there is significant penetration in the market. • GM expects to launch a "Supercruise" feature, enabling automation on highways, by 2020 [note: target date of 2017 announced by GM CEO in October, 2014 at ITS World Congress in Detroit] • GM considers digital maps and vehicle-to-vehicle and vehicle-to-infrastructure essential to vehicle automation
Alex Mitchell World Economic Forum	• The key question is what will get consumers excited about vehicle automation? • During the Annual Meeting in Davos (with invitees from Google, Cisco, Qualcomm, State Farm insurance and the United States Senate) the Governors agreed on the societal benefits of self-driving vehicles and how to shape a collective industry message. They agreed to create a roadmap, recognizing that the self-driving vehicle process will take many years of iterative planning and multi-stakeholder dialogue. • The process of deliberations by the World Economic Forum starts in 2014 with an overview of prerequisites for automated vehicles, discussing current challenges and prerequisites required to make automated vehicles a reality to be used as communication material for regulators • They will then develop an automated vehicle readiness index that measures each country's readiness for automated vehicles and allows derivation of implications and facilitates discussions on improvement areas • Finally, a roadmap for automated vehicles is developed based on scenarios of a 2025+ world, identifying key steps per stakeholder and highlighting significant changes (e.g., shift in risk landscape/liabilities). • The final output will be a report on Self-Driving Vehicle Readiness

Summary of Panel Discussion

Key points made with speakers in the subsequent panel discussion:

- Auto rental companies of today may evolve to shared vehicle companies in the future
- There is no consumer market for a driverless car (it will be shared vehicle companies)
- The only reason for a taxi driver today is to reposition the vehicle. The passenger could easily drive the car if it repositioned automatically without a driver.
- The car will be just a computer on wheels. The vehicle will just be another device between information and the consumer, just like a smartphone, tablet computer or television.
- A possible market force is companies providing (and requiring the use of) automated vehicles to employees for their commute.

Summary of sub-breakouts on the business case for automated vehicles

PVAC Sub-breakout 1

- SAE level 3 is simply a path to get to levels 4 and 5 and may not be mass produced
- 38% of Americans don't have driver's licenses but cost is a barrier to serving them
- In early stages the value might simply be a "gee whiz factor"
- A driver may rely on automation and not be able to reengage
- Automation could lead to driving skills degradation
- Weather will continue to be a barrier to full automation and dependence on automation
- SAE Level 5 will generate hundreds of new business cases

PVAC Sub-breakout 2

- SAE levels 3 and 4 are more relevant
- SAE level 3 is a necessary step to level 4
- Building a consumer and business case is difficult at this time
- Working while commuting appears to be a key incentive
- Getting younger drivers into safer driving cars is a key incentive, pre-teens
- Management of the interface between the driver and the vehicle is a major issue – how to manage the warnings, directions, and roles of the vehicle and the driver
- Expensive sensors in the front of a car may increase the cost of remaining accidents
- The car could become an alternative to mass/air travel
- Automated cars could displace other means of travel
- Shared vehicles are very compelling with SAE level 4
- Driverless cars will greatly increase vehicle miles traveled

IV. Key Results

- While the popular press has perhaps over-hyped the promise of vehicle automation in terms of how soon it will occur and how completely automated vehicles will be, nevertheless significant progress is being made and the roadmap for launch of incremental automation features is well underway.
- The markets of shared vehicles and vehicle automation appear to be tightly connected, automation enabling viable business cases for shared vehicles and shared vehicle demand incentivizing the development and launch of automated vehicles
- There was a strong consensus that SAE level 3 Conditional Automation, where the driver is expected to respond appropriately to a request to intervene, may not have a solid business case because it is unrealistic to depend on such behavior. However, SAE level 3 is a necessary step on the way to levels 4 and 5.
- Legislation and regulation were identified as challenges but workable over time
- Myriad changes to society will occur with the advent of widespread vehicle automation

V. Next Steps

The commercial launch of automated vehicles, whether conditional automation, high automation, or full automation (SAE levels 3, 4 and 5, respectively) is in process today. While the traditional auto manufacturers are proceeding with incremental feature launches, some members of the industry are advocating more aggressive, disruptive product launches. If a particular new automated vehicle offering resonates with a highly attractive business proposition, there could be acceleration of the market, at least for that particular vehicle configuration and application. On the other hand, if market impedances dominate, such as legislative or regulatory issues, then even incremental launches of increasingly more highly automated vehicles could be delayed.

There is significant value in anticipating disruptive market launches due to perhaps currently unknown value propositions as well as anticipating barriers that may unexpectedly deter adoption. Further work would be valuable in investigating these factors in more detail.

Appendix H: Technology Roadmap, Maturity and Performance: Operational Requirements for Vehicle-Road Automation Systems and Components Proceedings

Session and Proceedings Organizers: Jim Misener (Qualcomm), Ching-Yao Chan (University of California Berkeley PATH), John Estrada (Dering & Estrada)*, Ryan Gerdes (Utah State University), Yeganeh Mashayekh (University of Pennsylvania, previously of Carnegie Mellon University), Sudharson Sundararajan (Booz Allen Hamilton), Wei-Bin Zhang (University of California Berkeley PATH)*

I. Session Focus and Goals

Visions and concepts for automated vehicle-highway systems have as a common element the expectation of transformative personal and/or societal benefits. The performance of these future systems is fundamentally based on robust, reliable, safe and efficient technologies.

This session sought to examine state-of-the-art developments and technologies that will lend the needed support to realize the benefits of vehicle-road automation. Through presentations and group discussions, the session explored the needed performance, reliability and resiliency of these technologies using the framework of a broad and representative set of visions or states of automated vehicle-highway systems and their intended benefits.

II. Day 1 Presentations and Discussion

Day 1 Technology Briefings

- John Absmeier, Delphi, Vehicle Electric Electronic Architectures
- Roger Berg, Denso, Key Hardware and Software Components

Speakers	Summary of Presentation, Question and Answer Session
John Absmeier, Delphi *Vehicle Electric Electronic Architectures*	• The technologies in vehicles have trickled in as safety features. It took some safety features a long time to get adopted by industry, but the pace is accelerating driven by the convergence of consumer and automotive electronics. There is an appetite among the public for being connected. • In Europe, a 5-star Euro New Car Assessment Programme requires automatic emergency braking. • There are high levels of smartphone penetration in general, and it is nearly universal in younger demographics. • Automated driving technologies require a broad set of capabilities: o Sensors and perception systems—radar, LiDAR, vision, ultrasonic. Some exist and some are still in demo phases o Computing platforms and control systems—with the digital explosion, we have seen mobile device penetration driving the availability of high computing platforms o Electrical architecture and network management—how to make things talk to each other efficiently and effectively o Vehicle connectivity – inside the vehicle and externally o User experience – human machine interaction and interface. o Off board (cloud support and services)

	◦ Functional safety and security—when you open them up to exposure and connect. • NHTSA has defined 4 levels of automation ◦ Level 2 and level 3 are what we're starting to see coming into reality. No one has shown a level 4 car yet. We are still relying on humans to be part of this equation • Development stages ◦ Traffic jam assist – radar and vision ◦ Automated highway driving—production: 360-degree radar and GPS, demonstration ▪ Automated urban driving—production: 360-degree radar and vision, GPS. Demo:, V2X, HD vision, LiDAR • Autonomous vehicles will not be a huge game changer until we get to level 4, and we aren't there yet. For big societal effects like reduced congestion, car sharing, whole market impacts, need level 4. If you have one autonomous vehicle, you can replace six conventional vehicles at that point. • Historically, safety systems have reduced traffic fatalities by about 26%. Fatalities will eventually reduce with active safety systems and automation of more functions, but there are a lot of technology failures. There are going to be different and new types of accidents. When more of the control is ceded to the machine, there will be a tipping of the scales where humans will be less responsible. It is an ethical issue. • The bottom line: there are many challenges that still need to be overcome.
Roger Berg, DENSO *Key Hardware and Software Components and Connected Automation*	• Sample connected automated safety applications DENSO has developed and deployed include: ◦ Blind intersection warning—a demonstration video showed that if both vehicles are equipped, it is possible to see the results of a missed collision. ◦ Forward collision warning—a demonstration video showed a stopped vehicle on a curved path, in a situation which sensors may not handle as well as connectivity. ◦ Emergency vehicle advisor—most likely, AV sensors would not be able to sense the presence of emergency vehicle sensors. • DENSO has already had discussions with some OEMs potentially considering deployment of connectivity, whether or not there is a mandate. A wide spectrum of vehicles can benefit from this kind of technology. • For imminent crash avoidance, DSRC is almost surely necessary. • One of the important technologies to develop, maintain, and improve is driver status monitoring/sensory awareness. • As we look to the future and as society changes, it necessitates a holistic approach in how you sense, react, and actuate an intelligent vehicle system.

The session organizers provided attendees with the following set of deployment scenarios for small group discussion:

- A large western country decides that as of 2017 all newly manufactured vehicles must come with connected vehicle safety technology and that within a span of several phase-in years all cars must be equipped or retrofit with the capability.
- A small town (about 20,000) which is mostly a wealthy retirement community mandates that on a certain date that vehicles registered and traveling within the city limits must be autonomous vehicles.
- A medium sized country with virtually no oil or natural gas resources mandates that all trucks driving on the country's highways must utilize connectivity and automation to reduce the fuel consumption by x%.
- A large corporation in conjunction with a medium size city, about 400,000, launches a driverless taxi service to service the entire metropolitan area.
- A large state plans to modify its hot-lanes so that to incentivize AVs, for example, providing specific time windows for AVs and offering free or reduced access fees. Initially the incentives are limited by taking into account the market penetration but they are expected to expand over the next couple of years until they become 24/7 for AVs.
- A municipality decides to build a new light rail system using driverless vehicles. After successful launch of this new system, they intend to replace existing light rail transit routes with this technology.
- A large multi-national company decides to sell a NHTSA Level 3 vehicle in 2020.
- Like London Centre City, in 2018 a large US Metropolitan area decides to levy a use "tax" in a large section of the city during specific times and days (i.e. Green Zone). This tax is much less (or zero) for electric AVs.
- In 2015, a large US University with a large contiguous suburban campus allows only automated vehicles within a significant portion of its campus.
- A western military invests in automated technology mandating that 25% of its vehicles are automated by 2017.
- In 2016, a large Nationwide US package delivery service deploys platooning technology (NHTSA Level 1 or 2 and Connected Vehicle technology) to its cross country fleet and starts organizing platoons. It has previously received government approval to optimize the work rules surround platooning truck drives.
- In 2020, The US government mandates platooning trucks or automated vehicles only in the far left lane for large sections of the US interstate highway system.

Below is a summary of the technology challenges small group discussions for the three future states of automation selected by each of the small groups for discussion: (1) levels 3 and 4 automation, (2) mandated platooning, and (3) required connected capability.

- Levels 3 and 4 Automation:
    - <u>Core technology challenges</u>: Sensing and localization (HD maps and dynamic update); artificial intelligence; development of training software; integration with HMI; obstacle detection and avoidance in unanticipated scenarios (e.g., pedestrian on the highway); and passive and active safety designs for low-speed collision and coordination between systems).

- Operational environment and policy challenges: Standard service scenarios (or driver-by driver with training) and standard obstacles and reaction times the manufacturer is not responsible for (vehicle will minimize harm).
- Safety/reliability challenges: Skill decay and emergency/complex driving situations and license endorsement and retraining (simulator required?); development and testing of artificial intelligence software; complexity of system integration; data collection for validation; and redundancy systems.

- Platooning:
  - Core technology: Standardized networking; robust control (sensor noise/attacks/dropped packets); redundant systems (of same and different kind); and whole perception systems (especially urban perception) available for mass market.
  - Operational environment and policy challenges: Decision about whether to have continuous train or distinct platoons would be needed; sensitivity of sensing and communications to weather; willingness to pay for automation that is not available 100% of the time; developing joining and merging protocols; selection of an autonomous vs. human lead; and spacing to disband or transition.
  - Safety/reliability challenges: Which failure modes can be accommodated; tradeoff between safety and close spacing; adaptive maneuvering regimes for heterogeneous vehicles; and regulations to standardize when automation can be used.
- Required Connected Capability:
  - Core technology: Bandwidth and cognitive radio and notification (without display?)
  - Operational environment and policy challenges: Use of infrastructure or peer-to-peer; channel sensing to alert user of non-operational status or overload; defining functionality before radio technology, with consideration of what is being communicated; and consideration of how non-users can benefit.
  - Safety/reliability challenges: Understanding the effect of frequency on use.

III. Day 2 Presentations and Discussion

Day 2 Technology Briefings--Current and prospective programs to and test systems

- **Ching-Yao Chan**, University of California Berkeley PATH, Defining Systems and Test Scenarios
- **Jack Pokrzywa**, Society of Automotive Engineers, SAE Activities that Address Automotive Safety Integrity Level

Cross-Cutting Safety and Reliability Challenges:

- Whether there is Federal, state, OEM, and third-party testing is a regional preference that varies across countries.
- How to deal with aging and re-certification remains an open question.
- Model certification (aerospace level of testing/performance would be cost prohibitive). Testing and creating data on all possible scenarios would not be possible.
- An adversarial environment must be considered in the certification process: outline standard threats.
- Identified three degraded modes: fail safe, fail operational, and fail soft.

IV. Next Steps

- Complete a more on-depth "for the public good" use case approach
- Technology requirements for different operational policies
- Examine available versus necessary technologies
- Consider standards, certification, and regulation
- Fill gaps with research, investment and policy including: safety, robustness and reliability needs; innovation needed; and necessary policy steps

Appendix I: Road Infrastructure Needs of Connected-Automated Vehicles Proceedings

Session and Proceeding Organizers: Carl Andersen (FHWA), Paul Carlson (Texas A&M University), Valarie Kniss (U.S. DOT Volpe Center), Mohammad Poorsartep (Texas Transportation Institute)

I. Session Focus and Goals

This breakout session discussed how digital and physical infrastructure can support connected-automated vehicle technology. Day 1 focused on digital mapping of the infrastructure, including data needs, collection, availability, accuracy, and reliability of data. Day 2 was directed towards physical infrastructure advancements, such as enhanced roadway markings and intelligent traffic signals, in an environment where connected-automated vehicles are present.

The goal of the session was to better understand infrastructure and related data needs for connected and automated vehicles from different stakeholder and industry perspectives. We worked to identify opportunities and challenges in both digital and physical infrastructure as well as stimulate discussion about research needs, opportunities for innovation, and potential next steps to prepare infrastructure to be supportive and compatible with automated vehicles. The session focused on the following key questions:

- What is included in the definition of digital infrastructure and how does it support connected-automated vehicles? Is it more than just mapping? What infrastructure related data are needed and how are they obtained? What are the hurdles faced in collecting and maintaining these data?
- How can asset management practices today be integrated with data needs for connected-automated vehicles? What changes to existing practices would be required in order to support these vehicles?
- How can physical infrastructure advancements support connected-automated vehicles? How can vehicles use both physical and digital infrastructure in an integrated fashion?
- What technical and policy challenges exist when discussing digital and physical infrastructure on a national scale for connected-automated vehicles? What research is needed to address these challenges?

II. Day 1 Presentations and Discussion

Opening Remarks

The Tuesday breakout session discussed the concept of digital infrastructure and mapping in order to begin exploring the following questions:

- What elements are included in the definition of digital infrastructure?
- What data are needed and how will the data used by the vehicle? By the infrastructure?
- What is the government's role and responsibility in digital infrastructure and mapping?
- What is the relationship between industry and infrastructure owners and operators in an automated vehicle environment?

The presentations and open discussion covered the following topic areas:

- Data needs of various public and private stakeholders
- Data collection and processing
- Data accuracy and validation
- Performance and reliability
- Certification and standards for data
- Data ownership, coordination, dissemination and use
- Monitoring infrastructure changes and data updates

Panel/Presenter Summaries

The session started with a series of brief presentations followed by an open forum active debate on digital infrastructure needs to support connected-automated vehicles. Presentations focused on key topics, significant breakthroughs, opportunities, or challenges from the perspective of the speaker's respective industry in regards to infrastructure needs.

Speakers	Summary of Presentation, Question and Answer Session
Mr. Robert Dingess, President, Geospatial Transportation Mapping Association	Mr. Robert Dingess' presentation focused on the importance of holding conversations between the developers of automated vehicle technologies and entities responsible for overseeing transportation systems. This is particularly important for identifying the potential benefits of collecting geospatial transportation mapping data for both the needs of government and private industry. The following key issues were identified during the presentation: • **Tort issues serve as key institutional barriers to transportation innovations:** Tort liability protections seem to be slow in developing for automated vehicles. There are a variety of questions about where and how protections should be applied. • **Accurate base maps and V2I technologies create inherent tort risks:** State transportation agencies should be involved in discussions relative to the development of tort liability standards for automated vehicle base maps if data for these maps are being collected and provided by the States. Federal and State transportation agencies should be engaged in the discussion of standardizing the collection of base map data. There should also be a tort liability system that will cap issues related to when vehicles crash as a result of using these base maps. It is important to determine if the liability standards should be addressed at the state or federal level. • **State transportation agencies play a key role in V2I from a prioritization standpoint:** Entities involved in digital mapping and vehicle development should engage state agencies to begin discussing infrastructure needs. As State agencies better understand and grow more comfortable with these new technologies, they can be advocates for the changes potentially needed and help to identify roadblocks for automated vehicles. This could lead to a variety of options for overcoming roadblocks, including potentially certifying corridors and creating incentives for automation.

Mr. Alexander Klotz, Vice President Business and Technology Innovation, Continental AG	Mr. Alexander Klotz discussed the role of automated vehicles and connectivity in solving some of the key challenges in the world today. He described how automated vehicles can be used to reduce traffic collapse and commuting time, improve transportation efficiency, reduce C02 emissions, and reduce fuel consumption. The following key points were raised about connectivity and automated vehicles: • **Optimizing with connectivity:** With connectivity you can have optimization in a closed loop by reducing waste and increasing efficiency. This is important for automation since sitting in an automated vehicle while stuck in traffic congestion isn't necessarily desirable. Automation should be smart enough to be useful. • **Benefits to the driver:** With automation, a complete user experience is being envisioned (office = at home = one the go = in the car), something similar to the smartphone ecosystem. The major differences for automated vehicles is the information must come from a trusted source, the vehicle must be able to accommodate changes over the life of the vehicle (5+ years), and the environment must be non-distracting. • **Benefits of connectivity:** Connectivity can provide automated vehicles highly accurate and up-to-date digital maps for self-localization and environment interpretation, extended preview information to address physical limitations of in-car sensors, an extension of limited in-vehicle resources (such as processing power, memory), and highly accurate and validated data via crowd sourcing approach . It was noted that even though the vehicle is connected, the final decision on driving strategy remains with the vehicle, therefore the vehicle needs to recognize if connectivity has been lost and maintain functionality without connectivity. With connectivity, there are many questions facing industry such as: • How will we develop employee skills for the new world? • How will we secure data and intellectual property? • How will we connect suppliers, employees, and consumers to the cloud? • How will we monetize the data? Who owns the data? Who stores it? • How will we create consumer experience? • How would we change our business model for the automotive industry?
Mr. Ron Singh, Chief of Surveys and Geometronics Manager, Oregon Department of Transportation (ODOT)	Mr. Ron Singh's presentation detailed the Oregon Department of Transportation needs for maintaining day to day operations (engineering, planning, and maintenance functions) and the digital data that are being collected to support these needs. This was used as a launching point to begin discussing if there are other potential uses for the data beyond the DOTs needs. Digital infrastructure is supporting ODOT's needs through the following: • Collecting digital and imagery data provides a safer, smarter, less disruptive way to do surveying. ODOT can now measure many elements without having to go to the field. This information can also be overlaid with other in-house data. • 3D Design for highways, bridges, and tunnels allows for advanced visualization, clash and problem detection

	• Automated machine guidance and intelligent construction systems: ODOT can take digital 3D data and plug the information into construction equipment (graders, excavators, pavers) to perform automated construction functions. • Sophisticated asset management for informed decisions • Increased efficiency of operations of the highway system This was followed by a discussion of how these data could be relevant to connected and automated vehicles. Several key points about the data were highlighted, including: • The data are created, certified, and managed by expert qualified personnel: From an engineering perspective this is particularly important if you are creating engineering designs/plans from this information, this may be a different process than what is required for automated vehicles • The accuracy for 9,000 miles of highway is within a few tenths of a foot in order to determine where a lane line is located. Data can be extracted refined to within a few centimeters in locations where engineering quality accuracy is needed. • Once a highway project is completed, the as-built drawings are digitally generated and incorporated into the overall highway data, along with the original engineering design data and construction data.
Mr. Ogi Redzic, Vice President Connected Driving, HERE Nokia	Mr. Ogi Redzic described three key problems that HERE is working to solve within an automated vehicle environment including: 1) where exactly am I? 2) What lies ahead? and 3) How can I get there comfortably? HERE is delivering maps as a connected service, for both automated driving capabilities and consumer navigation, with multiple layers served from the platform to address these key problems. The solutions to these questions include: • **HD Map:** This map provides precise road and reference data, and is used for positioning for lateral and longitudinal control of vehicles on the road surface. HERE collaborated with OEMs on what was needed from a data and mapping perspective for automated vehicles. HERE is currently working on solutions for how maps should be updated when changes occur. • **Live Roads:** This addresses the need for accurate planning of vehicle control maneuvers beyond sensor visibility. Live Roads is in the early stages of development and uses machine learning, information from the vehicle controller area network (CAN) bus, and data feeds from smart infrastructure to let the vehicle and driver know what is happening on the road now. HERE is currently working on standardizing the way that real-time roadway information is collected. • **Humanized driving:** HERE is working ways to make automated driving more comfortable for humans. Automated functions will be customized for individual experience when information gathered from sensors (breaking force, etc.) is overlaid onto the other maps. It is envisioned that eventually there will be individual maps for each vehicle type as well as individual drivers.

<table>
<tr><td></td><td>HERE is able to generate high precision maps due to the business of global scale and far-reaching capabilities for both automation and navigation functions. HERE is already providing traffic data to State agencies so that they can access traffic pattern data; however no sensor data incorporated into the humanized driving elements are being shared.</td></tr>
<tr><td>Mr. Serge Lambermont, Technical Director Automated Driving, Delphi Electronics & Safety</td><td>Mr. Serge Lambermont's presentation provided an overview of digital infrastructure and other interrelated activities supporting the development of automated vehicles, including the role that suppliers play in this environment. Two key areas were highlighted: 1) understanding the strengths and weaknesses of technology today and how digital infrastructure can be used to further support automated vehicles; and 2) exploring the uses and impacts of digital infrastructure in adjacent markets today that could then lead to further developments in automated vehicles. The following are points raised from these two key areas:

Human versus computer vision/radar recognition
<ul>
<li>Assuming that human are perfect drivers, how would this compare to machines? A quadrant diagram with computer recognition and human recognition on each axis was presented to demonstrate the strengths and weaknesses of both groups (see figure below)</li>
<li>Maps for localization and lane markings can immediately address areas where human recognition is strong but machine recognition is lacking, including road boundary detection, curved lane marking, crossing lane marking, adjacent lane marking, stop lines and lines for lane marking.</li>
<li>ADAS maps can address stop lines, traffic lights, traffic signs, and lines for lane markings, which can be recognized fairly consistently by machines but this could provide improvement and redundancy.</li>
<li>V2I functionality can be used for determining trajectories of other vehicles/objects, and reading traffic lights and traffic signs, as well as V2V redundancy.</li>
</ul>
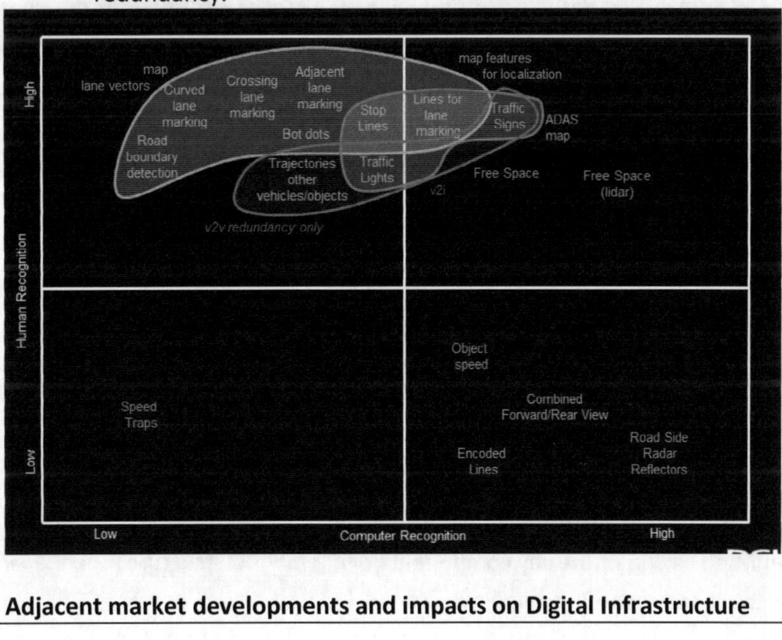

Adjacent market developments and impacts on Digital Infrastructure</td></tr>
</table>

	• **Passenger Car Market & Commercial Vehicles:** This market continues to be driven by improvements to active safety and New Car Assessment Program changes. This includes automated braking, detection systems, active collision avoidance, automated highway lane change. This market uses digital infrastructure for ADAS maps, V2X, functional safety and cybersecurity. • **Low Speed Autonomous – no steering wheel:** This is a somewhat new market; it is envisioned that these vehicle types would primarily be used for retirement, airports, entertainment, work campus, care free city centers, manufacturing, etc. There is more confidence today for programs at low speeds; we are seeing a large number of low speed prototyping vehicles emerging. These vehicles use maps but are also very dependent on the sensor suite. Digital infrastructure is used for localization. • **Tech enabled car share and rideshare:** This market is already using digital infrastructure to a small extent as information is being transferred to the cloud and the vehicle. This requires dynamic digital infrastructure – optimal booking-routing, real-time cloud based updates, new physical infrastructure, and call on demand. • **Personal Rapid Transit:** This includes smaller vehicles (2/3 wheel, limited width), coming to market for fuel efficiency and individual mobility purposes. These vehicles may need to share lanes with regular vehicles and could impact the use of digital infrastructure.

Summary of Group Discussion – Digital Infrastructure

The group discussion session was designed as an open forum providing attendees and speakers an opportunity to talk about topics related to both digital and physical infrastructure needs for automated vehicles. There were two primary goals of the session: (1) to identify challenges and opportunities, both technical and policy, where potential infrastructure innovations and changes are needed to support connected-automated vehicles; and (2) to discuss research needs and next steps to prepare infrastructure to be support of and compatible with connected-automated vehicles.

The following questions were used to initiate discussion about digital infrastructure:

- What is included in the definition of digital infrastructure and how does it support connected-automated vehicles? Is it more than just mapping?
- What infrastructure related data are needed and how are they obtained? What are the hurdles faced in collecting and maintaining these data?

Below is a summary of key points discussed during the session.

Mapping

Mapping was identified as one of the key components of digital infrastructure. The following general questions and comments were raised regarding mapping:

- Are mapping data critical to automated vehicles? It seems that there needs to be a series of systems or components that address various failure modes, but is mapping one of the critical components? Or can automated vehicles rely on other systems and components without the use of maps?
- Mapping already plays a role in vehicles today. As an example, some vehicles today use map data for things such as roadway slope and feed this information into the transmission/engine control to reduce fuel consumption. It seems like a natural progression that the map data would continue to be enhanced and used for automated vehicles.
- While it is important to discuss the quality and accuracy of maps, there could be a point where you cannot trust the map data. For example, what if overnight there is a landslide or earthquake that changes the structure of the roadway, etc.?
  - It could be the most accurate map and was fully certified, but the final decision needs to fall with the vehicle; can't design/implement a system that goes crazy with any one fault
  - From a system perspective, if vehicle sensors are in conflict with the map, the vehicle still needs to act safely - this is the real challenge – needs to be a system approach, not just looking at individual functioning components

Data

Data are needed to create both static and dynamic digital maps. One of the key points of the digital infrastructure discussion was the possibility that data could be "collected once, used often" for multiple purposes. States are collecting data today for various uses including asset management, performance evaluation and safety analysis. OEMs are developing highly precise maps for navigation and automation purposes.

Is there a possibility that data could be shared, creating benefits for all involved parties? How can we maximize the data for an automated vehicle environment as well as for State DOT purposes? Session participants identified benefits and opportunities, as well as potential issues and challenges, in sharing data. Research needs and potential next steps were also discussed. The comments below are grouped as data needs; collecting and sharing data; standards and certification of data and maps; and roles and responsibilities.

Data Needs

- There seems to be conflicting responses from OEMs regarding information and data element needs for high precision maps – for example, one OEM needs a particular attribute; but another OEM needs something different. Several possible explanations were discussed.
  - Each company is beginning to figure out what they need but since they are using different technology suites and components; depending on where emphasis is placed, there are varying needs; still a lot of learning ahead of the auto manufacturers
  - Technologies may need to be a compromise of affordability, development and production timelines, different business models/perspectives, varying technology architectures, functionality (for example, automation for parking vs automation for highways, depends on what auto manufacturers are targeting today) – all of these elements could explain the discrepancy between opinions
- With a lack of consistency or definition of data needs for automated vehicles, it seems to be difficult to create a baseline for knowing if data being collected by State DOTs can be used for automated vehicles or not.

- We need to be careful to not prescribe a solution for digital mapping and collecting data, which could eliminate the competitive market. What if Company X wants to develop its own global solution beyond the United States and the individual states?
    - It was noted that States are collecting data for asset management purposes; not exclusively for providing this information to automated vehicles.
    - It's important not to prescribe "x, y, z", but sharing the data gives States an opportunity to monetize what they already have for their own purposes and to potentially help independent mapping companies and auto manufacturers maximize their efforts, and also potentially accelerate the timeframe for deployment of automated vehicles
- What are some of the potential options for monetizing data and/or maps?
    - Several DOTs have offered data to auto manufacturers for their use; it would be helpful to know how this could be packaged for monetization. It was noted that monetizing is not the State focus, just a secondary benefit after collecting data for other purposes.
    - Is there a possibility of having a data exchange? Data being collected by vehicles could be extremely valuable for planners and designers to better understand the vehicle dynamics on particular roadway segments. This potentially results in better roadway design, improved environment for automated vehicles, and improved base maps.
    - Where will be the tipping point when the data collected by the vehicles becomes more valuable than the data being offered today (potentially a function of quality and volume)?

Collecting and Sharing Data

- Some state agencies are collecting and using digital data for asset management, engineering, and construction purposes. Often times the LiDAR data are stored in a point cloud and is only extracted when needed. The question was raised as to how it is decided from a DOT perspective what elements are extracted from the LiDAR point cloud and at what resolution. For example, if you wanted to pull out cross slope, there are a hundred different ways to calculate cross slope and there is no one standard definition. The response was "it depends" – on use, location, etc. For example, during construction, depends on the accuracy needed for the automated construction vehicles. Depending on how the data were collected, they can be extracted at various accuracy levels. Some DOTs have extraction capabilities in-house.
- Agencies indicated that they are potentially willing to adjust data collection practices if they know what is needed; the first step is continuing to discuss with OEMs the data and mapping needs for automated vehicles. Is it possible to find consensus from a DOT perspective on what elements should be collected? But then from all OEMs perspectives as well? Will there be enough commonality to start generating a national base map?
- The data discussion seems to be primarily focused on roadway infrastructure, such as lane markings, but vehicles also need to know where they cannot go (terrain information beyond roadway surface). It is important to know what is outside the traveling lane – are State DOTs currently collecting this information?
- Although the exact data elements needed to create digital maps for automation may be unknown at this point, collecting data in a consistent manner is essential. If the same elements are included in every data set and have been collected using the same standards, this could be valuable to OEMs since then they will be able to look at what is available and determine what they can use from the data.

- If these data can be shared, will there be one repository for the data? It seems difficult to gather information from each individual state, county, and city, etc.
- The asset management data being collected by States are typically not collected or updated often, although this is somewhat dependent on the particular asset. States typically don't revisit for several years in areas that are primarily static.
    - How often are updates required in static base maps for automated vehicles? What requires an update?
    - What are the potential issues associated with collecting data every couple of years? Would States be required to collect data more often if they are not required for asset management purposes?
    - In some states, the goal is to generate digital as-built drawings, including utilities, for all new infrastructure projects (roadways, structure, etc.). The data for these as-builts will be of engineering quality/actual survey data with very high accuracy.
- Are the data being collected by State DOTs complete and accurate enough for automated vehicles? One State DOT explained that each part of the LiDAR and point cloud data is collected at a different accuracy level; so it depends on the data collection source and use. Each source has its own accuracy levels, then each data point also has an accuracy level
- Gathering construction and/or work zone information for all roadways to generate dynamic maps could be a complex process
    - Google indicated that they have received work zone data from Caltrans; but Caltrans only controls about 8 – 10 % of the roads in CA; this leaves a huge part of the State network as unknown
    - How could you put together a complete data set of all roadways? States are limited in how much they can control. Typically State, County and Local agencies have a difficult time with coordinating roadwork as it is. Would companies collecting data need to gather from all roadway agencies on an individual basis – local, county, state, etc.?
    - What types of information would need to be identified and how often are updates required? Simply that a particular portion of the roadway is under construction, or are updates required every day as traffic barriers move, etc.?

Standards and Certification

- What standards are DOTs using now for data collection and map generation?
    - Is there a standard format that can be utilized so that everyone can access? Is there a way that this information can be consolidated now so that OEMs can review all at the same time, then can compare internally to their data sets and determine what can be used? This way OEMs can adapt their own practices internally or figure out what they need to extract and then provide input back to the DOT if specific elements should be changed without having to share in a forum that may impact competitive practices.
    - HERE uses a single specification – Navigation Data Standard (NDS) – sees value in using a global approach.
- Who is going to certify the base map?
    - If a base map is generated (either by State DOTs or another source of map generators) and it is handed off to the client customer (for example, potentially an auto manufacturer), how can accuracy be guaranteed? What standard is used?

  - Who is going to define "certification" for data and maps?
  - What about the liability associated with using a generated base map? How does the liability vary from state to state and how does this impact certification?
- Could State agencies certify base maps based on their collected data (self-certification)? Or what would their role be in the certification process if they are providing the data?
  - One State DOT explained that from a Civil Engineering perspective, the digital 3D as-built data/drawings/models are certified by a license surveyor because this information will be used for subsequent engineering design
  - Another DOT's initial reaction to this questions was that they would not want to be responsible for certifying the data; but then stated they would be willing to at least have a conversation about what this could possibly entail
  - Each state has their own nuances for surveying and certification, this is going to be a significant challenge
  - Is certification necessary? Would the mapping companies and auto manufacturers be willing to take the data as is? In what cases?
- How are data sets certified for accuracy? What role does tolerance play in the certification process?
  - What does the term accuracy mean for data sets? Are just the raw data certified or are the base maps as well? What is the process?
  - What is the tolerance that you can have for error? Tolerance for error is very different for navigation vs automated driving..... consider the example of 1 meter, if you are dropped off 1 meter from your intended destination, not a big deal, but if you are 1 meter over the centerline, a much bigger deal
  - How do you keep maps up to date? Do the maps need to be re-certified every time they are updated?
  - It is difficult to gauge the interest level from mapping companies in being the party responsible for certifying base maps

Roles and Responsibilities

- What is the problem that automated vehicles are trying to solve? Addressing this question could help to identify the roles and responsibilities associated with digital infrastructure, mapping and data. Also, it seems there are a number of issues that automated vehicles could address, but whose decision is it to decide? For example, is the primary purpose of automated vehicles for efficiency? Consumer focus / convenience? Depending on the problem, does it matter where the data come from?
  - One of the key concepts for automation is being able to have high capacity and high efficiency roadways. This is an important concept for State DOTs in making the roadway more efficient.
  - If you have poor mapping, this results in low speeds because automated vehicles will resort to using sensors, decreasing efficiency
  - You can use map data for strategic capacity decisions if the automated vehicles are connected.
- It's important to think about automated vehicles and its relationship to digital and physical infrastructure, but also need to be thinking about automated vehicles within a system
  - System manager – who can play the role of the system manager? See the ISO 17427 document – discusses the roles/responsibilities for cooperative ITS

    - Thinking about digital and physical infrastructure but also governance and the role of certification, someone needs to write the standards and keep up to date, these are new areas and new language for the industry
    - Look at the European mandate for Cooperative ITS as an example, can have one manager for one jurisdiction or one state
  - What kind of coordination function is currently available to discuss these efforts of sharing data, creating maps, and developing digital infrastructure? Do we need all 50 States or can it be narrowed to a couple of key players? Who has the ability to make decisions on standards, data formats, etc.?
    - Need a champion to fund the effort and develop various business models for funding – think more creatively than just having States offering the data to OEMs
    - Then can use these models to solve the issues noted above – efficiency, supply vs demand, capacity, etc.
    - Is it as simple as just developing a large database that each of the States can input their data into?
    - The Geospatial Transportation Mapping Association uses webinars, annual conference, to provide concrete example of how the data sets are valuable to others and agencies, beyond just for its intended purpose
- Can it be required that authorities responsible for roadways have digital data on new infrastructure projects as they are updating the roadways? Understandably there aren't always data available on legacy roadway segments, but why aren't DOTs mapping with new construction?
- Even if the data come from other sources (non-DOT), shouldn't the DOTs still be involved in some way with industry collecting data and generating maps? What are the benefits and/or challenges to the DOT involvement?
  - For example, an independent company maps the data and the following day construction begins on the roadway, how will the independent collector know?
  - In general there is worry that the maps will not be updated as often as they should unless the vehicles are constantly updating the mapping data themselves using sensor data
  - If an independent company is collecting data for mapping, what is the possibility that States could use these data for engineering purposes, as they are doing now?
  - The States may not be able to use these data for engineering purposes because they are likely required to be certified by a professional from that particular State, for example, in Oregon, the data are certified by a professional surveyor.
- Are State DOTs prepared to invest more in the data collection if the cost is much higher than what is required today? If it turned out that the vehicles needed much more precision or more content, then would need to be discussed, but if it were something that could be implemented now and solves an issue we face today, it would seem that States may consider this, for example, if we can show that the additional data collection would improve efficiency, then States may be willing to begin collecting if results can be demonstrated in the near term

Summary of Day 1 Results

The discussion focused on identifying digital infrastructure needs and exploring the current status of digital infrastructure technology and activities today in support of automated vehicles. This includes the collection, availability, accuracy, and reliability of digital infrastructure data, as well as identification of policy, institutional and regulatory barriers to deployment of automated vehicles.

States today are collecting digital data today for a variety of needs, including asset management, safety analysis. Private companies are also collecting digital data to generate high definition maps for navigation and vehicle automation purposes. The question was raised about the potential opportunity to share data between public and private entities to support the deployment of automated vehicles. This includes developing partnerships for sharing data for mapping and navigation purposes as well as sharing data gathered by the vehicles to improve the transportation system.

Several potential benefits and opportunities were identified for sharing data:

- Gathering data from multiple states could allow for increased production of static basemaps leading to faster deployment of automated vehicles.
- Data collected from the vehicle and sensors could be extremely valuable for planners and designers to help better under the vehicle dynamics on roadway particular segments. This would result in better roadway design and an improved driving environment by potentially reducing the number challenging scenarios for automated vehicles.
- Data gathered from vehicle sensors could be used for asset management, specifically for pavement and performance management.
- If vehicles could report back lane marking information (quality, missing, etc.) this could be helpful to agencies responsible for roadway maintenance, especially if the roadway maintenance function is contracted out to a third party.

The following issues were identified when discussing digital infrastructure needs for automated vehicles:

- **Data Needs:** The data elements and related attributes needed by map developers and OEMs to support automated vehicles have not yet been articulated to State agencies. There also seems to be a lack of consensus from OEMs about what may be required. It is unknown if this is a general lack of interest in sharing the data or if the technology is still immature. Technologies may need to be a compromise of affordability, development and production timelines, different business models/perspectives, varying technology architectures, targeted functionality (i.e. automation for parking vs automation for highways) – all of these elements could explain the discrepancy between opinions.
- **Data Formats and Accuracy:** Data collected by State agencies contain various elements with different formats and accuracy levels depending on the how the data are collected and how they will be used (engineering design, asset management, safety analysis, etc.). Each data collection source has its own accuracy level, and each point cloud data element also has an accuracy level. One State agency has accuracy levels for 9,000 miles of highway within a few tenths of a foot in order to determine where a lane line is located. Data can be extracted refined to within a few centimeters in locations where engineering quality accuracy is needed.
- **Global Standards:** Industry may be more interested in data collected by States if it was collected using a global standard. Some companies are seeing value in taking a global approach and using standards such as NDS. Federal and State transportation agencies should be engaged in the discussion of standardizing the collection of basemap data. This would require collaborating with industry to identify infrastructure data needs to determine if existing data standards are robust enough for automated vehicles.
- **Data Retrieval:** The logistics of each OEM accessing data from 50 individual states could be challenging. Industry may be more interested in accessing data from states if it could be retrieved from one location rather than approaching each state. This may require implementing a national data

repository which raises the question of who would be responsible for funding and maintaining the database.

- **Certification:** It is unknown what certification entails for data and automated vehicles. Will the raw data be certified, will the post-processing functions be certified, or will the completed basemaps receive certification? If the data are certified for the day it is collected, what happens when changes are made to the roadway? If the data are being shared by States, would the State be responsible for certification or would a third party need to be involved? A State agency collecting data explained that the data are created, certified, and managed by expert qualified personnel: From an engineering perspective this is particularly important if you are creating engineering designs/plans from this information, this may be a different process than what is required for automated vehicles
- **Data Updates:** Static basemaps provide precise road and reference data, and are used for positioning for lateral and longitudinal control of vehicles on the road surface. The mapping industry is currently working on solutions for how maps should be updated when changes occur. States do not update their asset management information often; the map is primarily static. However, for newly completed construction projects, some states are digitally generating as-built drawings and incorporating this information into the overall highway data, along with the original engineering design data and construction data
- **Dynamic Events:** Providing updates to vehicles for roadway changes is a significant challenge. This includes any elements that are not static, such as work zones. One map developer is in the early stages of developing these types of maps, which use machine learning, information from the vehicle CAN Bus, and data feeds from smart infrastructure to let the vehicle and driver know what is happening on the road now. HERE is currently working on standardizing the way that real-time roadway information is collected.
- **Tort and Liability:** Accurate basemaps and V2I technologies create inherent tort risks. State transportation agencies should be involved in discussions relative to the development of tort liability standards for automated vehicle basemaps if data for these maps are being collected and provided by the States. There should also be a tort liability system that will cap issues related to when vehicles crash as a result of using these basemaps. It is important to determine if the liability standards should be addressed at the state or federal level.
- **Optimizing with connectivity:** With connectivity you can have optimization in a closed loop by reducing waste and increasing efficiency. This is important for automation since sitting in an automated vehicle while stuck in traffic congestion isn't necessarily desirable. Automation should be smart enough to be useful.
- **Benefits of connectivity:** Connectivity can provide automated vehicles highly accurate and up-to-date digital maps for self-localization and environment interpretation, extended preview information to address physical limitations of in-car sensors, an extension of limited in-vehicle resources (such as processing power, memory), and highly accurate and validated data via a crowd sourcing approach. It was noted that even though the vehicle is connected, the final decision on driving strategy remains with the vehicle, therefore the vehicle needs to recognize if connectivity has been lost and maintain functionality without connectivity.

III. Day 2 Presentations and Discussion

Opening Remarks

The session on Day 2 discussed the various types of physical infrastructure and traffic control devices changes needed to support connected and automated vehicles. This included current practices today as well as ongoing research in these areas. This was followed by a discussion on DOT infrastructure data collection activities, including the current federal requirements for asset management, performance management and safety improvement. A key point in the discussion was being able to understand what fundamental data elements are being collected today by infrastructure owners and operators, how that information is being collected, and what potential benefits as well as hurdles that may exist in collecting additional data to support connected and automated vehicles.

The presentations and open discussion explored the following topics:

- Intelligent traffic signals, including Signal Phase and Timing (SPaT)
- Roadway markings
- Roadway lighting
- Asset management practices and data needs from a State DOT perspective
- Data collection technologies

Panel/Presenter Summaries

The session started with a series of brief presentations followed by an open forum active debate on physical infrastructure needs to support connected-automated vehicles. Presentations focused on key topics, significant breakthroughs, opportunities, or challenges from the perspective of the speaker's respective industry in regards to infrastructure needs.

Speakers	Summary of Presentation
Dr. Kevin Balke, Research Engineer, Texas A&M Transportation Institute	Dr. Kevin Balke's presentation focused on the use of signalized intersections within a Connected-Automated vehicle environment. He highlighted several key challenges and issues with infrastructure that may need more consideration as these vehicles are deployed, such as: • **Fixed versus flexible roadway use:** The physical geometry of the roadway is fixed but the use of the roadway can vary by time of day, day of week, or special occasion, for example, reversible lanes or turn prohibitions. Actively managing the transportation system could also vary movements at a particular intersection, either on cycle by cycle basis or short-term basis. We need the ability to communicate to vehicles the current use of infrastructure at any specific point in time. • **Timeliness of communications:** Connected and automated vehicles can be a rich data source for traffic management purposes. Being able to bring this info into a traffic management system and process info in a timely fashion is challenging. It takes time to collect, process, make a control decision, and disseminate information. We currently don't have a good way of handling this information and turning the information around quickly to send back out to vehicles. Also, the information disseminated to the driver and the infrastructure must be in agreement.

	• **Prioritization of command and control decisions:** There are potentially a large number of applications that can be put into the vehicle. From a traffic management standpoint, not all decisions may have equal precedence. Who decides on priority of command and control of applications? • **Rapid evolution of technical advances and maintaining equipment:** If automated vehicles rely on particular infrastructure systems, who is responsible for deploying and maintaining these systems? What if new systems are needed that don't exist today (i.e. cloud, subscriber services, etc.)? What is the government's role in this new infrastructure? The speed at which government can deploy new infrastructure seems incredibly slow as compared to the evolution and growth of automated vehicles. • **Maturity and consistency of standards:** Multiple government entities typically oversee roadways and sometimes there is no one "overseer" of the operations. The National Transportation Communications for ITS Protocol (NTCIP) process can be long; need to be able to think through the process of what we really need to standardize vs not standardize. The idea of fast versus slow adopters of new standards also needs to be considered.
Dr. Paul Carlson, Division Head Traffic Operations & Roadway Safety, Texas A&M Transportation Institute	Dr. Paul Carlson discussed roadway markings being used today and the challenges being faced by auto manufacturers and Tier 1 suppliers when developing technology for automated vehicles. This was followed by an outline of research being conducted to produce more reliable and improved detection of pavement markings for machine vision. The following topics were identified in the presentation: • **Roadway markings today:** Varying types of roadway markings exist today; there are multiple acceptable variations of marking types within the MUTCD (for example, there are dozens of contrast marking types). Each State is able to develop guidelines and preferences for marking types as long as they fit within the MUTCD parameters. Automation technologies have a difficult time detecting certain types of markings consistently. It also makes vehicle technologies more complex as so many different markings require detection. • **Lab research for improved markings:** There is ongoing research on how pavement markings can more reliably detected by camera systems for day/night contrast, glare saturation, wet nighttime conditions using detection systems such as infrared systems and cameras and high bit cameras. • **Updating standards and specifications:** Organizations, such as the American Society for Testing and Materials (ASTM) and American Association of State Highway and Transportation Officials (AASHTO), are willing to put together different standards and specifications on infrastructure to support AVs if we can determine what enhancements are needed from an automated vehicle perspective. • **From research to implementation:** Some new research has been identified that could be incorporated as part of a new pavement marking specification. Opportunities are also being sought to try some of the new research ideas in the field, particularly with existing automated technologies today

Dr. Ronald Gibbons, Director of the Center for Infrastructure Based Safety Systems, Virginia Tech Transportation Institute	Dr. Ron Gibbons' presentation detailed the trends in roadway lighting today, including the implementation of Solid State technologies and the resulting move towards Lighting Control systems, Master Lighting plans, Adaptive Lighting depending on the needs of the environment, and managing lighting levels as an asset. This was followed with discussion on how connected and automated vehicle can use lighting and the associated potential issues, such as: • **Lighting on demand:** Would provide lighting for a vehicle as it is needed when is passes. Currently being investigated are issues such as how much light is needed and how much light in front of a vehicle is someone comfortable with, can you use this level of comfort and lighting to control speed, and what are the visual performance issues. On demand lighting would need to consider pedestrians (potentially needing an X to I link) or the public (homes in close proximity to the road experiencing flashing lights) • **Lighting requirements for automated vehicles:** It is likely that lighting is needed for up to level 3 automation. But even if lighting were not required for automated vehicles, you would need to consider pedestrians, homeowners feeling of security, driver comfort and situational awareness, and legacy vehicles still needing lighting.
Mr. Jim Misener, Independent Consultant	Mr. Jim Misener's discussion focused on the importance of connectivity in an automated vehicle environment. The discussion began by exploring the three potential future vehicle worlds: 1) the connected vehicle world; 2) the independent automated vehicle world; and 3) the intersection of these two worlds – the connected-automated world. It is clear that SpaT and DSRC are important for the connected and connected-automated world, but how relevant are they for an independent automated world? The following key points were addressed as they relate to communications and connectivity for automated vehicles: • **Use of DSRC by automated vehicles:** DSRC broadcasts communication from short range with low latency; it provides a way for vehicles to know if there are other vehicles in close proximity, by broadcasting "Here I Am" messages. This is primarily used for safety purposes. But if there were no connected vehicles, or no DSRC, could automated vehicles still broadcast information about the vehicle? Or could it communicate in some other way with other vehicles and/or infrastructure? Need to be aware of other physical means beyond DSRC to deliver messages, such as LTE, 5G, etc., in the case that DSRC is not deployed or if there is not ubiquitous DSRC coverage during early deployment • **Applicability of message sets:** In addition to the physical means of communication and communication layers, Need to start thinking about how to separate out automated vehicles from DSRC and consider which message sets are applicable to these vehicles. SAE is already doing some work in this area. An automated vehicle may or may not have DSRC but it still needs to communicate to receive/update maps – how can you get this information via other communication methods (cellular, etc.) and what needs to change in the message sets? It is also important to think about the applications for automated vehicles and applications without vehicles (V2P) and what messages are needed, how they differ, and what means of communication could be used for each.

Mr. Chris Harris, P.E., Civil Engineering Manager, Maintenance Division, Tennessee Department of Transportation (TNDOT)	Mr. Chris Harris described TNDOT's efforts in collecting and using digital asset data over the past several years. The main focus was on relating TNDOT's requirements for asset management to how these efforts relate to connected and automated vehicles. The following issues were raised as they relate to State agencies and the future of asset management: • **Minimum data requirements for Asset Management vs Automated Vehicles:** For asset management, States typically collect data based on minimum requirements to maintain assets and for budgeting maintenance costs. This includes things such as square feet of signage, linear feet of guardrail, etc. It does not require elements to be located in space. TNDOT has chosen to map their assets in space but this may not be typical for all States. • **Collecting and maintaining data:** If State DOTs have a role in providing data to automated vehicles, then specific data requirements for accuracy, timeliness and consistency are needed. • **Funding for data activities:** Currently there is no dedicated federal funding for collecting data. Existing federal funding can be used for data collection, but this becomes a trade-off for collecting data vs building and maintaining infrastructure. If State DOTs are expected to provide data to automated vehicles, are there opportunities for shared cost between States and industry?
Mr. Ray Mandli, President, Mandli Communications	Mr. Ray Mandli introduced Mandli Communications, which is a data collection company for various entities including State DOTs, airports, cities, and counties, and collects imagery, pavement data and 3D LiDAR information for a complete asset inventory. The discussion highlighted the following issues related to collecting and sharing data with automated vehicles: • **Complete asset inventory:** Mandli collects an enormous amount of rich, dense data including imagery, pavement condition data, and 3D LiDAR information to build a complete asset inventory for various entities, including State agencies, airports, cities, and counties. Assets include: clearances, drainage, edge types, intersections, intersection layouts, lane types, luminaires, medians, mile post markers, monuments, pavement messages (arrows on pavement), pedestrian crossings/ramps, power pedestals, sidewalks, sign assemblies, sign faces, signal assemblies, cabinets, poles, traffic lights, utility lines, concrete barrier walls, delineators, guardrails, lanes, paint striping, pavement surface area, rumble strips, shoulders , sign supports. • **Data collection using federal funding:** Many states pay for these data collection with federal funds. As a result, these are considered public domain data and are available via the Freedom of Information Act. The data are delivered to each state in their own format but Mandli maintains the data in the raw format so that at any point they can be reused or remodeled. • **Proprietary vs non-proprietary issues related to mapping:** It seems that basemaps should be non-proprietary and the same data can be used for all maps, but all basemaps need to be consistent across states. Data from Mandli are currently in the Open Drive format, but it is unknown if this is

	the format that is needed for automated vehicles. It will need to be determined what base elements are needed for automated vehicles, then industry can build off these elements and maps to create proprietary products.
Mr. Tony Tavares, Chief, Division of Maintenance, California Department of Transportation	Mr. Tony Tavares' presentation provided an overview of the Maintenance Program at Caltrans, including key elements of the asset management program, challenges that Caltrans faces today and opportunities for improvement in asset management for the future. The following key points were made about DOT maintenance and the possible impacts that this would have on automated vehicles: • **Aging of existing system:** 80% of California's State Highway System was built between 1959 and 1974. Many elements have exceeded their original design life and need replacement. Specifically, 33% of all the ITS/Transportation Management Systems elements are no longer functioning and need replacement. How should existing maintenance needs be balanced with limited resources and the possible introduction of new infrastructure elements? • **Limited resources with competing outside needs:** California is faced with the challenge of having limited resources with competing regulatory mandates, natural emergencies (earthquakes, floods, landslides, fires) and copper wire theft. • **Funding shortfalls:** In 2006, Caltrans infrastructure maintenance needs total \$3.6 billion, today the needs total \$8.2 billion. This year's operating budget is \$2.0B, only 25% of the total needs assessed. This makes it extremely difficult to maintain and preserve assets. Transportation funding in California is also highly decentralized, often resulting in a mine-versus-yours tug-of-war between state and local governments. How will funding for infrastructure elements needed for automated vehicles be addressed at the local, state and federal level? • **State Priorities:** California has several plans in place for maintaining their transportation system, including the "Fix it First" initiative, the California Transportation Infrastructure Priorities and MAP-21 for Asset Management. How can partnerships be developed between state and local transportation agencies and the federal government that yield transportation investments in a single system that reflects local, regional, and state priorities?

Summary of Facilitated Group Discussion – Physical Infrastructure

The group discussion was designed as an open forum for all attendees and speakers to talk about any topics related to digital or physical infrastructure needs for automated vehicles. Below is a summary of several key questions and comments received during the two hour discussion on physical infrastructure.

Physical Infrastructure Needs

- In a recent field meeting, there was discussion between State DOTs and OEMs regarding infrastructure requirements for automated vehicles, specifically pavement striping. There were varying opinions from

OEMs about striping requirements. The session participants discussed possible reasons for this difference in opinion and noted the following points:

- It seems there are just different approaches for seeing and sensing the road, the OEMs appear to each have a different solution due to the competitive nature of the environment
- During a plenary session earlier in the week, there were comments about how no markings, signs, traffic signals, roadway lighting are needed at all for automated vehicles

- Systemically, what are the most problematic infrastructure elements for automated vehicles? What could be adjusted by agencies? What could agencies do to make things friendlier for automation?
  - Lane markings are important
    - When maps are built for the first time, need good lane markings – mapping the lane markings using highly precise equipment – is it impossible to get high precision lane level marking maps unless the lanes are marked in the first place?
    - Currently not using vehicle sensor data to update that maps/lane markings; primarily using sensors in the vehicle to report back data on roadway incidents
    - If vehicles would report back lane marking information (quality, missing, etc.) this could be helpful to agencies responsible for roadway maintenance, especially if they have contracted out roadway maintenance to a 3rd party
    - Once the maps/lane markings have been established using highly precise equipment, can the lane markings go away and the vehicles use data fusion through maps and sensor data?
    - Pavement marking installers now use sensors to be able to reapply markings exactly as they were placed previously
  - General redundancy is very important, for example, need to know if we have missed a sign if someone is walking in front of the sign and blocking the vision
  - Positioning at road edge is difficult and expensive; challenge for longitudinal positioning. Trying to use other furniture of the road for positioning, this is why knowing the exact location of signs is helpful in order to position the vehicle
- There was a general belief from OEMs that they have to live with what infrastructure is in place now and make systems work with that infrastructure. During the session, there was a breakthrough in discussions between OEMs and State DOTs. Infrastructure owners/operators expressed willingness to making changes to the infrastructure to accommodate automated vehicles if there is some guidance as to what should be done. The discussion revealed that there are many things that "could" be done since there are still so many unknowns in the automated technology field. The following questions were identified to begin thinking about potential changes:
  - Are there any changes absolutely required?
  - How are the changes prioritized?
  - Where would the biggest benefits be? State agencies don't necessarily have a good understanding of the automation technologies and systems and where the biggest benefits in infrastructure changes could come from.
  - All of these questions need to consider timeline and cost, as well as societal, policy and trust issues
  - If you wait to make any infrastructure changes and allow automated vehicle technologies to be developed based on infrastructure in place today, then the systems will be based on potentially dated infrastructure, not enhanced/advanced infrastructure. This is essentially a chicken and egg dilemma. If you base technology on infrastructure today, can it be adapted to "smarter"

infrastructure? Or do you make changes to infrastructure today in anticipation of automated vehicles?
    - How do infrastructure requirements vary based on the level of automation? What do we need today instead of getting caught up in what may be needed in 2050?
    - Can changes be phased? Possibly start with just highways, develop different kinds of markings to support automated vehicles, then can slowly implement technologies on all other roadways
- We heard during a plenary sessions that it is easier for vehicles to detect circular signs as compared to rectangular signs. We don't understand why this is true. Again, this reiterates that State agencies need a better understanding of infrastructure needs for automated vehicles and machine vision.
  - Currently there are so many variations between signs that the number of algorithms and "if-then" statements is almost endless for automated vehicles
  - Meaning of signs is also sometimes different depending upon the context, so perception is important, the vehicle isn't "thinking" like a driver would be. Also need to remember that signs are currently designs for human comprehension – for example, there are 2 signs for curves, a sloping curve and a 90 degree turn. There are rules for when each of these signs are applied, but not information is given to the driver about the radius, banking etc. of the curve because this would be too much info and it's not helpful to humans. There are limits to the amount of info on a sign for humans, but there is potentially no limit for vehicles. So does the composition of signs change all together? QRs, etc.
  - What is the first step in providing more consistency between states in terms of signage? What if there were easier things that could be done, such as putting a boarder around the sign or changing the color of the signs, what about different materials for the signs?
  - Chicken and egg issue again – Do we have sensors read what exists today or do we spend huge amounts of money to replace signage that is better for automated vehicles?
  - There are changes being proposed to the MUTCD that could add complexity to the roadway system, for example, infrastructure components for shared roadways between cyclists/motorist and as well as colored lane pavements (red for BRTs, purple for electronic tolling, etc.). How difficult is it for vehicles to see cyclists in the roadway?
  - Bicycle and pedestrian detection does exist today – fairly easy to detect both
  - It is harder for vehicles to detect "Generic objects" – objects that are separated from the road and not classified, but we know that vehicles can hit them, for example, a deer jumping in the road
  - What is difficult for machine vision for pedestrians is determining the intention of the pedestrian – for example, if a person is standing next to the road and gives physical cues, such as turning his head or raising his hand, a driver would understand that this means they are going to be crossing the street, technologies today aren't really able to detect these subtle cues

Engagement and Collaboration

- Agencies need to be more engaged than they are now in the automated vehicle-infrastructure discussion, even if it is difficult to gage where the OEMs are going.
    - It was surprising to some that there hasn't been more interaction to date between States regarding automated vehicles.
    - Some DOTs are currently working with Tier 1 suppliers and auto manufacturers and have set up roadway test sections.

  - Although some DOTs are engaged with the auto industry, there is no forum or means of communication between the DOTs not involved with the OEMs to share knowledge.
- The American Traffic Safety Services Association represents companies that manufacturer pavement markings. They have invited auto manufacturers to meetings to talk about what is needed from pavement marking so that this information could be relayed to Congress, but could not get companies to attend.
  - This could indicate that companies are still in the early stages of development;
  - Could also indicate a lack of interest - that companies have taken the approach that they will use what they have since there is no way to guarantee that these markings would always be available

Maintenance and Reliability

- While setting up the vehicle demonstrations for the symposium, there was concern about the faded pavement markings and bot dots used in CA and the ability for the vehicle technologies to work in these conditions. In the end, the demonstration was successful, which leads to the following questions and conclusions:
  - More study is needed on what contrast relationship is needed for pavement markings
  - Ongoing research is trying to identify where the vehicle systems fail and where they succeed, but more research is needed to determine what can be done with the infrastructure to improve reliability
- During construction, pavement markings are removed and replaced by temporary markings several times before the roadway is completed. This if often very confusing to drivers due to the lines not being removed completely. The same confusion may occur with machine vision.
  - Need to be able to remove and replace the pavement markings in the form/fashion that is not confusing to drivers and vehicles
  - Need to have better standards on removing and replacing pavement markings during construction
- Maintenance activities for maintaining pavement, for example sealing cracks in asphalt, can create a spaghetti mess of lines on the roadway. Has there been any research on what this does to the vehicle's machine vision? A mapping company noted that using a laser crack measurement system, you can map to a millimeter of a crack. And although it is sometimes difficult to tell the different between asphalt and concrete using the sensing equipment, they have not encountered issues with the asphalt sealing.
- There has been discussion about automated vehicles potentially increasing VMT by 5 – 7%. Considering there is always a natural increase in VMT, how will this proposed additional level of increase impact the pavement lifecycle?
  - How do you prioritize and determine what to focus on first? How will performance measures change with automated vehicles? Always a function of limited funds and constrained resources.
  - Should the transportation funding model be revised? How will it need to be revised with the onset of automated vehicle? Is there a relationship between the VMT model and automated vehicles?

Standards

- There was a lot of discussion by OEMs and map generators about not having roadway marking standards in place. However, there are standards currently in place for roadway markings; but there are typically various options that meet the standards. The same is true for traffic signals – some are horizontal, some are vertical, etc.

- Are the standards too flexible (bot dots, contrast markings, etc.)? Does the MUTCD give too many options and choices to States?
- Are the current standards strong enough to support automated vehicles?
- Are we not consistent in our application of the standards?
- Is it more of a maintenance issue after the markings have been in place? What are performance requirements and expectations from OEMs about infrastructure?

- In 1991, Congress required that FHWA establish minimum retro-reflectivity standards for pavement markings. FHWA has been working on this since 1992 and a standard still does not exist today.
  - Even if you set the standard as low as possible, wouldn't the result be that every marking would have this standard? Would this standardization be beneficial to an automated vehicle environment?
  - Why has there been no move to create a standard? Is more research needed? Is there politically not enough pressure?
  - OEMs did testify before Congress that "good markings" were needed – what does "good markings" mean?
  - Will there be a day when human driving (non-automated vehicles) is prohibited? Will there always be a shared roadway use between automated and non-automated vehicles? It seems that there may be a need for improved roadway markings for the near and extended future. (Look at the FAA example – have drones and Cessna sharing the same airspace, need to have criteria that have multiple vehicle types (motorcycles, manually driven vehicles, automated vehicles, etc.))

Summary of Day 2 Results

It has been assumed by many that automated vehicle technologies need to adapt to the physical roadway infrastructure in place today, but automated vehicles many be a big enough change in the transportation environment to warrant changes in infrastructure in the short-term as well as the future.

A major point in the discussion was that State agencies are open to considering changes to the infrastructure if the changes that are needed to facilitate the development and deployment of automated vehicles can be identified. More collaborative discussion is needed between infrastructure owners/operators and industry to determine infrastructure needs, priorities, benefits, and the associated timeline and funding requirements for these changes. Several key questions were raised: 1) How do infrastructure requirements vary based on the level of automation?; 2) What do we need today versus what is needed in the longer term to support automation?; and 3) How do you find a balance between developing new standards for infrastructure versus waiting for automated vehicles to be more advanced?

State agencies and OEMs have engaged in preliminary discussions on what is needed from physical infrastructure with limited results. There seems to be no consensus from OEMs, which could be explained by several factors: 1) automation technologies are still in the early stages of development; 2) OEMs are developing different technologies and approaches for seeing and sensing the road; 3) technologies are using what is available today so that there is less reliance on infrastructure; 4) sharing needs in a public forum could compromise competitive advantage.

Below is a list of issues and challenges identified for physical infrastructure as related to automated vehicles that could help to start discussions between industry and the public sector.

- OEMs testified before Congress that "good markings" were needed for automated vehicles, but what constitutes good pavement markings is not clear. It is unknown if this relates to materials, styles, maintenance, etc.
- There are multiple acceptable variations of marking types within the MUTCD (for example, there are dozens of contrast marking types). Each State is able to develop guidance and preferences for marking types as long as they fit within the MUTCD minimum guidelines. Automation technologies have a difficult time detecting certain types of markings consistently. A possible area of research could be to identify which marking types are most easily and reliably detected by automated vehicles and then begin the process of updating the MUTCD to reflect these findings. Also it should be determine if there is a minimum level of maintenance required for pavement markings.
- Temporary pavement markings are applied in work zones, often multiple times during a course of a project. The markings need to be able to be removed and replaced in a form/fashion that is not confusing to drivers and vehicles. Improved standards on temporary pavement markings may be necessary.
- It is unknown if roadway lighting is needed for all automated vehicle technologies and to what level of automation is lighting necessary.
- Signs are currently designs for human comprehension – for example, there are 2 signs for curves, a sloping curve and a 90 degree turn. There are guidelines for when each of these signs are applied, but no specific information is given to the driver about the curve – radius, banking etc. There are limits to the amount of information provided on signs that is useful to humans, but there is potentially no limit for vehicles. How would this change the composition of signs? For example, would signs include QRs, etc.? Or is the solution simpler, such as change the sign shape to round instead of rectangular or outlining the sign in a particular color?
- Many States currently face funding shortfalls for deployment and maintenance of existing infrastructure elements. These limited resources are also competing with outside needs, such as regulatory mandates and natural emergencies (earthquakes, floods, etc.). Would infrastructure elements required for automated vehicles be placed at the same priority level as other infrastructure elements? How could this impact the deployment of automated vehicles?
- Automated vehicles may rely on particular infrastructure systems or new systems that do not exist today (i.e. cloud infrastructure, subscriber services, infrastructure supporting connectivity, etc.). Would it be the role of the government to deploy and maintain this infrastructure? The speed at which government can deploy new infrastructure seems incredibly slow as compared to the evolution and growth of automated vehicles.
- Multiple government entities typically oversee roadways and sometimes there is no one "overseer" of the operations. Need to be able to think through the process of what really needs to be standardized vs non-standardized as the standards process can be time consuming. The idea of fast versus slow adopters of new standards also needs to be considered.

Several of these issues are being addressed through continued research, for example, there is research being conducted on how pavement markings can more reliably detected by a variety of camera and infrared systems for day/night contrast, glare saturation, and wet nighttime conditions. Opportunities are also being sought to try some of the new research ideas in the field, particularly with existing automated technologies today. Organizations, such as ASTM and AASHTO, are also willing to put together different standards and specifications on infrastructure to support AVs if we can determine what enhancements are needed from an automated vehicle perspective.

The next step is establishing more forums for discussion between industry and infrastructure owners and operators to begin exploring two key questions:

1. Is there an interest by industry to be engaged in these discussions? Or are technology developers satisfied with continuing to utilize existing infrastructure for future technology development?
2. If there are infrastructure changes that could help to improve the environment for automated vehicles, what are they?

IV. Key Results

The following are key points taken from both days of discussion on physical and digital infrastructure.

- States are collecting digital data for various purposes, including asset management, performance evaluation and safety analysis. Industry is also collecting digital data to generate high definition maps for navigation and vehicle automation purposes. There may be an opportunity to "gather once, use often" and share data to support the deployment of automated vehicles and achieve benefits for both the public and private sectors.
- Additional discussion is needed between industry and states to identify what data and related requirements are necessary to support connected-automated vehicles in order to determine if leveraging State collected data are viable.
- Currently States are collecting data in using various collection standards and data formats; standardization of these processes may be required to share data or generate a common base map.
- States are responsible for deploying and maintaining traffic control devices. States are willing to make changes to infrastructure in support of automated vehicles if guidance is provided. This requires more discussion with industry and automated vehicle developed to determine requirements and priorities for physical infrastructure.
- Implementation of traffic control devices can vary state by state. Today's guidelines, such as the MUTCD, provide several alternatives for implementation; some of which may be more challenging than others for automated vehicles. Revision to standards and guidelines may be necessary.
- Traffic control devices are designed for human comprehension; multiple alternatives pose a greater challenge for automated vehicles.
- Low cost infrastructure changes could be implemented now; other changes can be factored into long term planning

V. Next Steps

The discussion of infrastructure needs seems to be only in the beginning stages. The following next steps were proposed in order to begin discussing what infrastructure changes may be required for both digital and physical infrastructure in order to facilitate the deployment of automated vehicles.

- Hold workshops at the national and state level to gather requirements from developers of automated vehicles
- Assess impacts of requirements on digital infrastructure data collection and maintenance of physical infrastructure
- Develop recommendations and guidance to share across all states to accelerate deployment of automated vehicles

Appendix J: Friday Ancillary Workshop—Envisioning Automated Vehicles within the Built Environment: 2020, 2035, 2050 Proceedings

Shannon Sanders McDonald (Southern Illinois University Carbondale), Caroline Rodier (Urban Land Use and Transportation Center, University of California at Davis), Kati Rubinyi (Civic Projects, Inc.), Ramses Madou (Stanford University), Marco Anderson (Southern California Association of Governments), Reuben M. Juster (University of Maryland College Park), Dimitris Milakis (Delft University of Technology), Susan Shaheen (University of California Berkeley), Ray Traynor (San Diego Association of Government), Elliot Martin (University of Berkeley California)

I. Session Focus and Goals

This ancillary all-day workshop presented the opportunity to focus on the policy and built environment issues related to automated vehicle use. The workshop committee focused the event on assisting MPOs in their efforts to address these new technologies. The workshop was financially supported by: University of California Davis, National Center for Sustainable Transportation, Southern California Association of Governments, ARUP, Kimley Horn, Fehr & Peers, and the National Center for Intermodal Transportation.

Steve Shladover started the day with an overview of the symposium, and then in order for the workshop attendees to have a base of knowledge several handouts were distributed titled *Technology Application Pathways to Commercialization*. In the interest of addressing the built environment three generic design sites were created: *Streets and Roadway, Neighborhood and District* and *Regional* for use in the hand-on workshop. Seven scenarios were developed and participants had the ability to choose the scenario group in which they wanted to take part. 110 participants took part in some or all of the day's agenda, representing a wide set of backgrounds and disciplines. Nine scenario groups formed each with a committee member and one or two facilitators to enable the discussions and drawings.

In the interactive workshop, participants discussed and envisioned the built environment impacted by automated vehicle technologies. Small, multi-disciplinary teams of experts combining knowledge of a wide range of fields - city planning, infrastructure and architecture, car design, engineering, software and systems – collaborated on six different chosen scenarios. Each team started with all three design sites (streets/roadway, neighborhood/district and regional scales) described in words, images, maps and diagrams. Scenarios were developed with a time-horizon and presence of specified types of vehicles with varying degrees of shared mobility, connected vehicle technology, and autonomous operation. The Scenario groups generated, through different interactive methods and writing/visualizing techniques, "after" scenarios in order to think through the challenges and benefits to our built environment that an autonomous mobility future can hold. Each team briefly presented their outputs at the end of the workshop for feedback and discussion with the wider set of participants.

II. Presentations and Discussion

Opening Remarks

- Dr. Steve Shladover, Symposium Organizer, University of California-Berkeley
    - Dr. Shladover presented an overview of the prior three days of the symposium, for those who did not attend it.
- Dr. Dan Fagnant, University of Utah (formerly University of Texas-Austin), A Convergence in Shared Mobility: Demand-responsive fully automated vehicles, for carsharing and ridesharing across Austin, Texas
    - Dr. Dan Fagnant started the workshop off with his research findings for Austin, Texas and the use of level 4 shared autonomous vehicles. Increasing degrees of vehicle automation will ultimately introduce profound impacts across transportation systems. New benefits will hopefully emerge in terms of safety, congestion, parking, and pollutant emissions, though the transport system may experience congestion costs from added VMT. While we do not know when fully automated vehicles (AVs) will be able to operate without any 'driver' or occupant on board, once this threshold is breached it will enable the introduction of new transportation modes, such as shared automated vehicle (SAV) systems, which is the focus of this dissertation work.
    - The Austin-based simulation results suggest that a fleet of SAVs could serve many if not all intra-urban trips with replacement rates of around 1 SAV per 9 conventional vehicles. However, in the process SAVs may generate around 8% new unoccupied/empty-vehicle miles of travel that would not exist if travelers were driving their own vehicles (because SAVs require re-positioning whereas privately-owned cars do not need re-positioning). With dynamic ride-sharing active (and allowing for up to 30% extra total service time) a replacement rate of 1 SAV per 11 conventional vehicles and just 5% empty-vehicle VMT may be achieved, with 5% of all served VMT shared (by distinct or independent traveling parties).
    - Of course, if rapid and inexpensive service is provided but ridesharing is limited, it is possible that this system could attract substantial numbers of formerly non-motorized and transit trips, resulting in excess VMT and worsened congestion, an outcome to be avoided from a public policy perspective.

- Bryant Walker Smith, University of South Carolina (formerly Stanford University), Government Regulation, Anticipation and Participation
    - Bryant Walker Smith, an internationally-leading academic, lawyer and engineer researching legal aspects of vehicle automation, provided an insightful presentation of how technology is adopted and the speed of which this has occurred in analogous contexts. He discussed what municipalities could not control: deployment of new technologies by automakers, federal motor vehicle standards, state vehicle codes and insurance laws, consumer preferences and the weather. But, municipalities can control: local infrastructure, vehicle fleets, transit systems, taxi cab regulations, local traffic rules, parking rules and land use. He provided a source for scholarly articles on topics relevant to MPOs on http://newlypossible.org.

Presenter Summaries: Technology Application Pathways to Commercialization

Two types of hard-copy handouts were provided in this section of the agenda: *Levels of Automation* and *Technology Application Pathways to Commercialization* handouts. The purpose of these handouts was to provide a common language and point of discussion for the participants, to ensure focus on the impact on the built environment rather than the trajectory of the technology. The Symposium provided 2 ½ days for technology-focused discussions and now the workshop participants were ready to discuss what it all might mean to our daily lives.

Speakers	Summary of Presentations, Question and Answer Session
Marco Anderson *Levels of Automobile Automation*	• A one page visual document of the NHTSA levels of automation was provided and discussed as a point of reference for the workshop attendees.
Reuben Juster *Levels of Transit Automation*	• A one page visual document was created, specifically for this workshop, for transit similar to the levels of automobile automation. An interactive presentation of the document took place.
Dimitris Milakis *Automobile Pathways*	• A three page brief on the state of the art of automated automobiles for four time periods: current, 2020, 2035, 2050. References and a glossary were also included.
Reuben Juster *Transit Pathways*	• A four page brief on the state of the art of automated automobiles for four time periods: current, 2020, 2035, 2050. References and a glossary were also included.
Dr. Susan Shaheen/Michael Galczynski *Taxi and Car Sharing Pathways*	• A four page brief on the state of the art of automated automobiles for four time periods: current, 2020, 2035, 2050. References and a glossary were also included.
Dr. Elliot W. Martin *Goods and Freight Pathways*	• A three page brief on the state of the art of automated automobiles for four time periods: current, 2020, 2035, 2050. References and a glossary were also included.

Presenter Summaries: Built Environment Design Workshop

The built environment handouts and presentations were based upon aspects of the built environment at three scales: streets and roadways, neighborhood and district and regional. These handouts provided information and specific context design drawings that could be used as beginnings for new urban design strategies.

Speakers	Summary of Presentations, Question and Answer Session
Reuben Juster *Streets and Roadways Design Site*	• Photographs, a street section and google earth views were provided addressing main street blocks and arterial blocks. These were provided at 8.5 x 11" for each participant, 11 x 17" for each group and larger format drawings around the room.
Ramses Madou *Neighborhood and District Design Site*	• Google views, street maps, and a land use document were provided for an example of a neighborhood condition. These were provided at 8.5 x 11" for each participant, 11 x 17" for each group and larger format drawings around the room.
Marco Anderson *Regional Design Site Information*	• Regional views of the San Diego area were provided. The arterial system, the highway networked system, low end communities and communities of concern, high-frequency local bus routes, transit priority project areas, smart growth concepts map, and the transit network, all SANDAG documents. • These were provided at 8.5 x 11" for each participant, 11 x 1 7" for each group and larger format drawings around the room

Dr. Carolyn Rodier *Scenarios*	• All of the scenarios were provided in 8.5 x 11" printed documents for each participant, please see description below.

Scenario Group Summaries

Attendees choose which scenario they were interested in exploring. The scenario was the organizing concept for the discussion. Each scenario had market penetration by time horizon identified, a scenario description, and assignment questions to help start or focus the conversation. The design sites handouts were used as appropriate to address the physical environment discussions involved with that particular scenario.

- Scenario One: Investment and Re-Design of the Freeway System
- Scenario Two: Complete Streets
- Scenario Three: Transit and Road User Fees
- Scenario Four: Commuter Rail Traffic
- Scenario Five: Parking
- Scenario Six: Mitigating Poor Performance
- Scenario Seven: Sustainable Mobility for All

All of the scenarios had full participation; six -eight attendees to a group with a committee member and at least one facilitator (these were architects and planners) except for scenarios three and four. So, scenarios three and four were not discussed at this workshop. A group of attendees wanted to define their own scenario and called themselves the *Revolutionaries*. So, we had nine scenario groups including the *Revolutionaries*. There were 80 participants in total in the scenario group part of the day's agenda.

Scenario Groups	Summary of Presentations, Question and Answer Session
One (Group A): Investment and Re-Design of the Freeway System Tom Faulders – Facilitator Kati Rubinyi – Committee Member	• Hot Lanes o Higher speed for automated vehicle lanes o Dedicated off-ramps o 100 mile per hour off ramps o Capital need is for ramps not flyovers o High-occupancy vehicle lanes converted to dedicated lanes o Dedicated lanes will assist with initial transition to the technology • Trucks o Segregate by trucks and speed o Truck only highways at night o Medians could be replaced by something else o Banking to deal with speed • Freeways o Variable speed lanes based on demand o Freeways become quieter and cleaner o Google maps to have an impact on traffic o Perception of efficiency vs real efficiency o Reclaim flying freeways o Public and Private Transit like Public and Private Road o Not just about efficiency and optimization but to connect to the

<table>
<tr><td></td><td>
physical environment
o Reclassifying freeways and changing design standards

• Ramps
o Could have super ramps steeply banked
o Shorter ramps – mini ramps integrated with arterials
o Business locating around ramps
o Ramps connected to parking structures that are also destination centers
o All structured parking can be converted to local distribution facilities

• Economics
o Can we build in advance of the market?
o Public Private partnerships around freight – could be key to paying for infrastructure
o Distributions centers
</td></tr>
<tr><td>One(Group B):
Investment and Re-Design of the Freeway System

Robert Alexander - Facilitator
Marco Anderson- Committee member</td><td>
• Background
o Increasing population 1/3 of the counties
o Decreasing population 1/3 of the counties
o Same in 1/3 of the counties
o Nature of work changes
o Time
o Still high Inequity
• Findings
o Increased VMT
o Increased Congestion
o Distributed Travel
o Peaks shallower and more distributed
o Increased density but not enough to keep up with demand
o Some traffic everywhere all the time

• 2030 Freeway network
o Change to VMT pricing mitigates “deadheading”
o High end level 4 personal ownership, higher speed, longer distance and off-peak travel
o Median households will still own cars, but not 2 cars only one, car sharing is used
o Infrastructure dynamic lane assignments, speeds etc.
o Public – private funding
o Funding at regional level
• 2050 + Network
o Variable lane directions, push lane speeds during congestion
o Less individual car ownership
o More different types of vehicles
o Private revenue advertising directed to in-vehicle experience
</td></tr>
<tr><td>Two(Group A):
Complete Streets

Tyler Folsom-Facilitator</td><td>
• Parking
o Frees significant areas for more interesting uses
o Remote parking
o Need to design drop-off areas as it replaces on street parking
o Allocate parking through pricing

• Bike lanes
</td></tr>
</table>

<table>
<tr><td></td><td>
<ul>
<li>o Recreational bikes need for separate bike lanes</li>
<li>o Commuting bikes can become automated and safely mix with rest of traffic.</li>
<li>o Virtual traffic control replaces physical</li>
<li>o Fewer physical signals, except for pedestrians</li>
</ul>
<ul>
<li>Different Street Types
<ul>
<li>o Fully automated</li>
<li>o Mixed</li>
<li>o Pedestrian/Bike</li>
<li>o Transit</li>
</ul></li>
<li>Less traffic calming apparatus</li>
<li>Narrow lanes, more right of way for pedestrians and parking</li>
<li>May need stations for drop-off at high demand destinations</li>
<li>Truck deliveries could happen at night, freeing roads for other uses during the day.</li>
<li>Retails space changes-the Internet may replace big box stores, with neighborhood showrooms for items one would want to handle.</li>
<li>More pedestrian spaces</li>
<li>Integration is now the key</li>
</ul>
Downside
<ul>
<li>Increased VMT? Congestion?</li>
<li>Physical health/activity</li>
<li>Sprawl</li>
<li>Social isolation</li>
</ul>
</td></tr>
<tr><td>Two(Group B):
Complete streets

Joe Kott, PhD, AICP - facilitator
Tim Siebert - Facilitator
Mariya Eggensperger
committee member</td><td>
<ul>
<li>Drawing of a proposed street section with automated vehicles: the team calculated that with the redesign of the street for multimodal automated vehicles, there would be 97% street use</li>
<li>Implication of Automation: automation encourages the blurring of physical road constrains & allows for a more relaxed convergence of public versus private space</li>
<li>Influences of Automation: Automation lessens the need for vehicular signage; automation fosters more case-by-case infrastructure project(s)</li>
<li>How does automation change the streetscape? Automation allows for simpler car lanes and more "human-scale" vehicles.</li>
<li>Drawing of car "pods:" Automation allows for the easy pickup and drop off of passengers. The roadway would accommodate this feature through a design called a <u>road "pod"</u> which is simply a strategically designed 'pocked' alcove intermittently placed along the road. (the graphic tries to demonstrate this idea of a road pod)</li>
<li>Drawing of the automated street "hierarchy:" A complete street not only factors in the safety of pedestrians, cyclists and vehicles, but also is organized and <u>multimodal</u>, accommodating several travel options <u>on one road</u>. Since automation will make vehicular movement more precise, the</li>
</ul>
</td></tr>
</table>

	need for a street 'hierarchy' is essential to right-of-way. • **Automation encourages density:** Since public parking with fully automated vehicles wouldn't be entirely required, space and land uses are repurposed for other functions such (as retail). • The advent of automation **encourages change in vehicular circulation,** the addition of different "transition and speed zones" would be required for the safe flow/function of automated vehicles • Automated vehicles' ability to perform door-to-door service would save people waiting time. • **Automated vehicles could change the way retail is done--** services/products are purchased online and automated vehicles would deliver. Since a driver would not need to be present to make the delivery, this cost is eliminated for businesses. • **Drawing of the hierarchical road with automated vehicles**: automation may encourage more structured control of roads yet also add a stronger level of "multi-mobility." • **Drawing of the automated vehicles' road**: automation will encourage "flexible and fixed road zones" • A Complete street should be defined relative to its context. • The issue of sharing streets or de-coupling modes of transport has a variety of benefits and detriments, given the situation. • Separate lanes can give full efficiency out of each mode of transport, however this gets extra complicated with the addition of a new mode - automated vehicles. Requires more infrastructure. • Automation has the potential to remove roadside parking from the street, which can be good/bad. When designed intelligently, cars can act as barriers for cyclists between the foot path and the road. • No street hierarchy: letting the streets sort themselves out. No lanes, no curbs, no delineation, no regulation. However this is the most difficult street for an automated vehicle to navigate through, given the large amount of unpredictable events and human social cues. But from a place-making point of view, could be the most desirable (e.g., in certain situations, see traditional European street, or busy market streets in India/Asia). • A hierarchical lane system makes crossing the street extra difficult and there will be more incentive to move into the third dimension via tunnels / overhead lanes / bridges etc. • These issues of infrastructure are best solved depending on the situation at hand. **Final notes:** • If we design smart cars we must also design smart streets • A complete street encourages street hierarchy to establish order and right-of-way, but does not hinder multi-mobility • Automated vehicles would simplify streetscape signage

<table>
<tr><td></td><td>

- Car "pods" are necessary on the road in order to allow for the safe pick up and drop off of passengers on the autonomous road.
- Digital presentation posted on workshop website
- Drawings Excluded, they will be posted on the workshop web site

</td></tr>
<tr><td>Five:
Parking

Gerry Tierney-Facilitator
Stan Young-Committee person</td><td>

Why do we need parking?

- Physicality of a vehicle – they need to go somewhere.
- Convenience : proximity to drop-off and pick-up
- Ownership? Parking protects an asset. Brought up discussion of owning vehicle or treating it as a service.
- Level of Automation allows for dislocation of building being served and parking.
- This may create (or move the congestion from the parking structure) congestion at the drop-off and pick-up areas, making the need to create a "White Zone" curb side.
- Congestion may result from too many vehicles going to & from the pick-up / drop-off
- This will result in curbside waiting / efficiencies
  - Just as in parking, there may be incentives and disincentives ($) for appropriate use
  - This may in turn bring up social equity issues as the curbside is traditionally public space, whereas parking is highly privatized
- Automated level 4 vehicles will park more efficiently
  - Less space required per vehicle for automated parking
- In the shared vehicle concept enabled by level 4, the aspect of the vehicle used as personal storage space is not addressed. Many people keep personal belongings in vehicle for convenience – sort of a personal toolbox.
  - Is this a major reason we own and drive private vehicles, or just a consequence? Will it hamper the sharing of level 4 automated vehicles?

Land Use and Public Realm Issue

- Will the elimination of public parking make downtown more attractive?
- Does the elimination of curb-side parking remove a barrier (a buffer zone) between pedestrians & the moving traffic? [Cars are known to travel slower on narrow streets lined with parked cars, if you remove the parked cars, will the traffic speed increase making it more hostile to pedestrians?]
- Where will autonomous level 4 vehicles go?
  - New, purpose-built parking structures elsewhere?
  - To existing parking [structures and curbside] made more efficient for automated vehicles
  - If parking in garage in suburban home not required, will this free up space (about 400 square feet per vehicle] for the in-law unit? (Grandma in the Garage]
  - Parking and land value will always be connected

</td></tr>
</table>

<table>
<tr><td></td><td>
<ul><li><ul><li><ul><li>Urban downtown too valuable for parking</li><li>Suburbia may have lower land values, providing more opportunities for parking</li></ul></li>
<li>Land values will dictate future use of vacated parking lot land [surface lots or parking structures]</li>
<li>However parking WILL still be required for parking vehicles – both level 4 and lower.<ul><li>Ground floors will/could become retail and upper levels remain parking.</li></ul></li></ul></li></ul>
</td></tr>
<tr><td>Six:
Mitigating Poor Performance

Brian O'Looney-facilitator
Ellen Greenberg - facilitator
Rueben Juster – committee member</td><td>
<ul>
<li>Automakers can be expected to market private ownership of autonomous vehicles</li>
<li>More AVs may lead to more mobility and access, which may lead to more sprawl</li>
<li>Congestion could worsen and downtowns will therefore suffer</li>
<li>People may choose live in their automated motorhomes, constantly cruising around popular activity centers, but causing even more traffic</li>
<li>Pedestrians will walk into the streets stopping automated vehicles and causing gridlock.</li>
<li>People's health will deteriorate since automated vehicles will cut down on walking</li>
<li>Many people's livelihoods will be disrupted (professional drivers and possibly mechanics)</li>
<li>To respond to this, governments can either be reactive by putting up fences on sidewalks so pedestrians don't disrupt traffic, or be proactive by charging vehicle automated vehicle passengers a VMT fee collected via a block by block EZ Pass system</li>
<li>Vehicle speeds in lively activity centers could be restricted to 5 mph to discourage vehicle use and increase safety while speeds of 125 mph would be allowed on dedicated infrastructure</li>
<li>In really high demand areas there would be requirements to reserve slots in traffic flow, like what is now done at high-capacity airports</li>
<li>Drawings Excluded, they will be posted on the workshop web site</li>
<li>Digital presentation posted on workshop website</li>
</ul>
</td></tr>
<tr><td>Seven:
Sustainable Mobility for All

Chris Roach-Facilitator
Ramses Madou – Committee member</td><td>
<ul>
<li>Policy Recommendations - Economics</li>
<li>Environmental</li>
<li>Social sustainability – access to mobility</li>
<li>Pockets of Density - networks</li>
</ul>
This group focused on sustainability issues that automated vehicles might help address.
<ul><li>The group set sustainability goals in the three primary sustainability areas. The goals were</li></ul>
</td></tr>
</table>

	• Environmental
	  ○ Decrease in energy and resource use
	  ○ Reclamation of space
	  ○ Decrease in pollution
	• Social
	  ○ Increase in mobility
	  ○ Safety
	  ○ Reduce spatial inequality
	  ○ Increase social contact/decrease waste personal time
	• Economic
	  ○ Sustainable transportation infrastructure funding
	  ○ Induce economic activity
	  ○ Reduce income inequality
	  ○ Demand management
	  ○ Lower mobility cost
	
	The group came up with policy and operational recommendation intended for a strong MPO audience.
	• Environmental
	  ○ Reclaim use of space
	  ○ Reduction in on street parking
	  ○ Reduce parking requirements for developments
	  ○ Dynamic virtual space allocation using virtual street infrastructure. E.g. ability to change lane directions at different times of day based on historic demand.
	  ○ New style building approach for vehicles. E.g. multi-tiered vertical approach, airport style multilane drop off and pick up space at first floor of buildings.
	• Social
	  ○ Reduce spatial inequality
	  ○ Reduce cost of mobility through more time efficient shared rides
	  ○ Food issues
	  ○ Subsidize automated food deliveries or roaming organic food stores to address food deserts.
	  ○ Demand based/flash mob farmers markets. E.g. farmers markets vendors upload their sales information and farmers can have automated stalls which bring data driven inventories to local farmers markets
	• Job access and mobility issues
	  ○ Provide access to automated vehicle systems, such as smart phones or carb side call buttons, to those who can't afford those items.
	• Safety
	  ○ Regulations/rules and monitoring of in vehicle behavior.
	  ○ Allow passengers to choose who they ride with.
	  ○ Social network like ratings of passengers. E.g. passengers can document creepy or other inappropriate behavior of other passengers.

<table>
<tr><td></td><td>
<ul>
<li>Access
<ul>
<li>Incentivization of supply to needy areas</li>
<li>Language and cultural diversity necessary in info and access points to automated vehicle systems</li>
</ul></li>
<li>Economic
<ul>
<li>Sustainable funding for transportation infrastructure</li>
<li>Pay as you use taxation, E.g. VMT or PMT (person miles traveled) built into business taxes for automated mobility service providers.</li>
<li>Will require tracking of vehicles positions, number of passengers and weight of cargo.</li>
</ul></li>
<li>Demand management
<ul>
<li>Complete system demand management. E.g. traffic management through real time dynamic road pricing on all roads with credits for those who share rides.</li>
<li>Dynamic road definitions. Central system control to alter road space usage based on historical and real time needs. E.g. change lane directions or shrink lane sizes for platoons of smaller vehicles.</li>
<li>Integrated mode management. Ensure that dynamic road definition include use by non-motorized vehicle modes.</li>
</ul></li>
<li>Lower mobility costs
<ul>
<li>Dynamic and incentivized sharing of vehicles</li>
</ul></li>
</ul>
</td></tr>
<tr><td>Eight:
Revolutionaries

Christer Lindstrom - Facilitator
Peter Mueller-Facilitator
Rachel Liu – Committee Member</td><td>
<ul>
<li>This group envisioned an entirely new “transit roadway” system that was solar powered, collected rainwater, multiple speed paths for varying purposes, had spaces for bikes and new automated mobility systems. This system was raised above the existing roadway, sheltering it and providing a new green roof surface for all manner of pedestrian activities.</li>
<li>This group had a lot of GRT & PRT industry members in it who may have felt under-represented by the proposed workshop topics – thus the call to anarchy which attracted other talents including two facilitators with no preconceived ideas about individual ownership of vehicles vs. public infrastructure.</li>
<li>Through a surprisingly sequential analysis of the Rapid Transit ideas presented by the GRT and PRT participants a concept evolved that is best represented by the sequence of sketches produced during the discussion.</li>
<li>The preliminary hypothesis of PRT cabins on a public elevated guideway system powered by the solar energy generated by its roof was retained very close to the images first provided by PRT industry activists but with the interesting hybrid twist of including the ability to carry privately owned autonomously guided electric velomobiles.</li>
<li>Laying out the groundwork for a gradual transition away from anything but ultralight vehicles in the cities.</li>
<li>Providing safe haven (protective envelope, public amenities, power for recharge, conveyance at regional and national speeds) for ultra-efficient (lightweight) ultra-personal (2- 4 seat public, public rental, or privately owned) vehicles (velomobile type cabins) when traveling longer distances at high speeds.</li>
</ul>
</td></tr>
</table>

- As illustrated by the sketches and notes; the anarcho-revolutionary group (by the end of the session they changed their name from anarchists to revolutionaries), a synthesis of ideas developed:
- A PRT (Personal Rapid Transit) Thesis based on ultralight above street conveyance system was examined by the group and embraced.
- The group zoomed out to look at how the system would work in different scales: A. First Last Mile ~20mph B. Local ~40mph C. Regional >100mph D. National (and beyond?) >200mph
- When the concepts of car minded participants were put on paper by the architect facilitator, differing socio-political views regarding ownership of transportation means showed themselves mid-session.
- Key concepts and values were examined to navigate towards a consensus based synthesis:
    - PRT is only efficient with ultralight cabins
    - Typically segregated from traffic for safety, electric vehicle cabins could descend to the street either;
    - At their own risk as an e-velomobile now
    - In the future with more safety after heavy vehicles a
    - In order to get the larger vehicles off the streets the cities should have major traffic re-routed to beltways with park and ride PRT system stations at outer edge.
    - Vehicle storage is a drawback of individual ownership 1. Mitigated by reliable summoning of public, rented, or private automated vehicles? 2. Vehicles used as storage of personal belongings?
- Key Values
    - Needs to be reliably on demand
    - Segregate people, bikes, ultra-light vehicles, autos, freight
    - Size grid to minimize last mile or allow cabins to drive on street.
    - Reduce or mitigate modal shifts (transfers) a. Piggybacking b. Roll on / roll off c. Carts for personal belongings
    - What does it look like?
    - High Density v Suburban
    - Social Changes in transit attitudes are needed a. Rental, public, car sharing b. Need privacy and security alternatives c. Time shifted work and shopping schedules d. Drone shopping
- Drawings Excluded, they will be posted on the workshop web site

III. Key Results

One of the most important outcomes from the Workshop was that it provided the forum for constructive multi-disciplinary interaction and collaboration, between disciplines that heretofore have been engaged in parallel discussions regarding vehicle automation. Having engineers, architects, planners, transportation professionals and technologists together discussing specifically focused scenarios allowed current problems to be explored and prospective and plausible future scenarios to be unpacked simultaneously from multiple angles. Policy, automated vehicles and physical planning were analyzed concurrently as a comprehensive system. Two divergent scenarios emerged from the day's discussions: one quite utopian in its vision, where automated vehicles could create an entirely new lifestyle and physical environment (Groups seven and eight) while the other assessed possibly-dystopian outcomes where sprawl increased (Group six). Group five (Parking) also highlighted genuine uncertainty, with visions of reducing parking on the one hand, but potentially also mechanisms that could increase parking and roads on the other. Scenario groups one and two examined the practical policy and potential changes in the physical environment. This link is a brief interview about the workshop that occurred the day of the workshop, http://viodi.com/2014/07/19/looking-at-the-impact-of-autonomous-vehicles-on-how-we-live/ – it is the organizers' hope that participants found the day thought-provoking and that it will stimulate further reflection about the future of vehicle automation.

IV. Next Steps

Generally-positive feedback was received from participants at this initial workshop; follow-up workshops are now in development. Informal networking and conversations between professionals and disciplines have started new research papers, proposals and research responses. A web site to post all the documents and presentations is currently under design, sponsored by the TRB APO20 committee: http://tinyurl.com/ldbgwzr. Anonymized feedback surveys and a list of participants will also be posted on this web site.

Note: This workshop came out of a discussion from last summer's 2013 TRB workshop on automated vehicles and was supported by the TRB Symposium committee as an important part of planning for automated vehicles in our built environment for the 2014 TRB/AUVSI Automated Vehicles Symposium. The committee would like to thank Jane Lappin, lead organizer of the Symposium for her strong and ongoing support for this endeavor.

Appendix K: List of Acronyms and Abbreviations

AASHTO	American Association of State Highway and Transportation Officials
ABS	Anti-lock braking system
ACC	Adaptive cruise control
ADAS	Advanced driver assistance systems
AERIS	Applications for the Environment: Real-time Information Synthesis
AFDC	Alternative Fuels Data Center
AMPO	Association of Metropolitan Planning Organizations
APA	American Planning Association
APM	Automated people mover
APTA	American Public Transportation Association
ARTS	Automated road transport system
ASCE	American Society of Civil Engineers
ASTM	American Society for Testing and Materials
ATN	Automated transit network
ATRA	Advanced Transit Association
ATRI	American Trucking Research Institute
ATTRI	Accessible Transportation Technologies Research Initiative
AUVSI	Association for Unmanned Vehicle Systems International
AV	Automated Vehicle
AVS 2014	Automated Vehicle Symposium 2014
BART	Bay Area Rapid Transit
CA	California
CACC	Cooperative Adaptive Cruise Control
CAN	Controller area network
CITS	Cooperative Intelligent Transportation Systems
CMOS	Complementary metal-oxide semiconductor
CTAA	Community Transportation Association of America
CSA	Compliance, Safety, Accountability
CV	Connected Vehicle
DARPA	Defense Advanced Research Projects Agency
DMV	Department of Motor Vehicles
DoD	Department of Defense
DSRC	Dedicated Short Range Communication
EU	European Union
FAA	Federal Aviation Administration
FDOT	Florida Department of Transportation
FHWA	Federal Highway Administration
FMCSA	Federal Motor Carrier Safety Administration
FTA	Federal Transit Administration
GM	General Motors
GPS	Global Positioning System
GRT	Group Rapid Transit
HD	High definition
HMI	Human Machine Interface
HMII	Human Machine Interface and Interaction
HDL	High definition LiDAR
ICM	Integrated Corridor Management

Ika	Institut für Kraftfahrzeuge
ISO	International Organization for Standardization
ITS	Intelligent Transportation Systems
ITS JPO	Intelligent Transportation Systems Joint Program Office
LiDAR	Light detection and ranging
MDOT	Michigan Department of Transportation
MOD	Mobility on demand
MPO	Metropolitan planning organization
MUTCD	Manual on Uniform Traffic Control Devices
LOS	Level of service
NASA	National Aeronautics and Space Administration
NDS	Navigation data standard
NFCTEC	National Fuel Cell Technology Evaluation Center
NHTSA	National Highway Traffic Safety Administration
NJ	New Jersey
NREL	National Renewable Energy Lab
NTCIP	The National Transportation Communications for ITS Protocol
ODOT	Oregon Department of Transportation
OEM	Original equipment manufacturer
ORAVS	On-Road Automated Vehicle Standards
PATH	Partners for Advanced Transportation Technology
PAVE	Princeton Advanced Vehicle Engineering Research Center
PCS	Pre-collision system
PMT	Person miles traveled
PRT	Personal rapid transit
PVAC	Personal Vehicle Automation Commercialization
ROI	Return on investment
SA	Situation awareness
SAE	Society of Automotive Engineers International
SANDAG	San Diego Association of Governments
SAV	Shared automated vehicle
SOV	Single-occupancy vehicle
TARDEC	United States Army Tank Automotive Research Development and Engineering Center
TMC	Traffic management center
TNDOT	Tennessee Department of Transportation
TOD	Transit-oriented development
TRB	Transportation Research Board
TSDC	Transportation Secure Data Center
U.S. DOE	U.S. Department of Energy
U.S. DOT	United States Department of Transportation
UC	University of California
V2I	Vehicle-to-infrastructure
V2V	Vehicle-to-vehicle
V2X	Vehicle—to- either vehicle or infrastructure
VMT	Vehicle miles traveled
VRA	Vehicle and Road Automation
WEF	World Economic Forum